ARCHITECTS, CONTRACTORS GUIDE TO CONSTRUCTI

			Section
DIVISION	1	GENERAL REQUIREMENTS 1A Equipment	1
DIVISION	2	SITE WORK & DEMOLITION 2A Quick Estimating	2
DIVISION	3	CONCRETE 3A Quick Estimating	3
DIVISION	4	MASONRY 4A Quick Estimating	4
DIVISION	5	METALS 5A Quick Estimating	5
DIVISION	6	WOOD & PLASTICS 6A Quick Estimating	6
DIVISION	7	THERMAL & MOISTURE PROTECTION 7A Quick Estimating	7
DIVISION	8	DOORS & WINDOWS 8A Quick Estimating	8
DIVISION	9	FINISHES	9
DIVISION	10	SPECIALTIES	10
DIVISION	11	EQUIPMENT	11
DIVISION	12	FURNISHINGS	12
DIVISION	13	SPECIAL CONSTRUCTION	13
DIVISION	14	CONVEYING SYSTEMS	14
DIVISION	15	MECHANICAL	15
DIVISION	16	ELECTRICAL	16
SQUARE FOOT COSTS			SF
STATISTICAL DATA			S
U.S. MEASUREMENTS, METRIC EQUIV'S, & AC&E ANNUAL INDEX			MM
ALPHABETICAL INDEX			AI

INTRODUCTION

EDITOR'S NOTE
2003

This annually published book is designed to give a uniform estimating and cost control system to the General Building Contractor. It contains a complete system to be used with or without computers. It also contains Quick Estimating sections for preliminary conceptual budget estimates by Architects, Engineers and Contractors. Square Foot Estimating is also included for preliminary estimates.

The Metropolitan Area concept is also used and gives the cost modifiers to use for the variations between Metropolitan Areas whose populations are in excess of 500,000 people. This encompasses over 50% of the industry. This book is published annually on the first of August to be historically accurate with the traditional May-July wage contract settlements and to be a true construction year estimating and cost guide.

The Rate of Inflation in the Construction Industry in 2001 was *3.0%*. Labor contributed a *4%* increase and materials rose *2%*.

The Wage Rate for Skilled Trades increased an average of *4%* in 2002. Wage rates will probably increase at a *4%* average next year.

The Material Rate increased *2%* in 2001. The main increases were in asphalt and gypsum products. The major decreases were in lumber, plywood and steel.

Construction Volume should increase in 2002. Housing will probably rise, and Industrial and Commercial Construction should stay flat. Highway and Heavy Construction should increase.

The Construction Industry again kept inflation under control in 2002. Some Materials should inflate at a slow pace, and Labor should again inflate slightly in 2003.

We are recommending using a *4%* increase in your estimates for work beyond July 1, 2003.

Don Roth
Editor

INTRODUCTION

Metropolitan Area Construction Cost Modifiers and Population (2000 Census) in the 50 largest Metropolitan Statistical Areas

Metropolitan Area	Population	Rank	Modifier	Metropolitan Area	Population	Rank	Modifier
1. Atlanta	4,112,198	11	90	26. Minneapolis, St. Paul	2,968,806	15	106
2. Austin	1,249,763	37	86	27. Nashville	1,231,311	38	91
3. Boston	5,819,100	7	118	28. New Orleans	1,337,726	34	89
4. Buffalo	1,170,111	42	114	29. New York	21,199,865	1	145
5. Charlotte	1,499,293	33	82	30. Norfolk	1,569,541	30	90
6. Chicago	9,157,540	3	114	31. Oklahoma City	1,083,346	48	89
7. Cincinnati	1,979,202	23	98	32. Orlando	1,644,561	27	94
8. Cleveland	2,945,831	16	110	33. Philadelphia	6,188,463	6	117
9. Columbus	1,540,157	32	100	34. Phoenix	3,251,876	14	98
10. Dallas, Ft Worth	5,221,801	9	85	35. Pittsburgh	2,358,695	21	107
11. Denver	2,581,506	19	96	36. Portland, OR	2,265,223	22	108
12. Detroit	5,456,428	8	114	37. Providence	1,188,613	39	107
13. Grand Rapids	1,088,514	47	100	38. Raleigh, Durham	1,187,941	40	82
14. Greensboro	1,251,509	36	82	39. Richmond	996,512	50	89
15. Hartford	1,183,110	41	108	40. Rochester, NY	1,098,201	46	110
16. Houston	4,669,571	10	87	41. Sacramento	1,796,857	24	116
17. Indianapolis	1,607,486	28	102	42. Salt Lake City	1,333,914	35	86
18. Jacksonville	1,100,491	45	91	43. San Antonio	1,592,383	29	87
19. Kansas City	1,776,062	25	100	44. San Diego	2,813,833	17	112
20. Las Vegas	1,563,282	31	108	45. San Francisco	7,039,362	5	116
21. Los Angeles	16,373,645	2	112	46. Seattle, Tacoma	3,554,760	13	108
22. Louisville	1,025,598	49	92	47. St. Louis	2,603,607	18	108
23. Memphis	1,135,614	43	94	48. Tampa, St. Pete.	2,395,997	20	96
24. Miami	3,876,380	12	88	49. Washington, Balt.	7,608,070	4	94
25. Milwaukee	1,689,572	26	106	50. West Palm Beach	1,131,184	44	93

(1) From Department of Commerce, 2000 Census
(2) AC&E Construction Cost Modifier - No Land Value Included

INTRODUCTION

HOW TO USE THIS BOOK

Labor Unit Columns
- Units *include* all taxable (vacation) and untaxable (welfare and pension) fringe benefits.
- Units *do not include* taxes and insurance on labor (approximately 35% added to labor cost). (Workers Compensation, Unemployment Compensation, and Employers FICA.)
- Units *do not include* general conditions and equipment (approx. 10%).
- Units *do not include* contractors' overhead and profit (approx. 10%).
- Units are Union or Government Minimum Area wages.

Material Unit Columns
- Units *do not include* general conditions and equipment (approx. 10%).
- Units *do not include* sales or use taxes (approx. 5%) of material cost.
- Units *do not include* contractors' overhead and profit (approx. 10%).
- Units are mean prices FOB job site.

Cost Unit Columns (Subcontractors)
- Cost is complete with labor, material, taxes, insurance and fees of a subcontractor. This also applies to Quick Estimating sections.
- Units *do not include* general contractors' overhead or profit.

Quick Estimating Sections - For Preliminary and Conceptual Estimating
- Includes all labor, material, general conditions, equipment, taxes, insurance and fees of contractors combined for items not covered in Cost Unit columns.

Construction Modifiers - For Metro Area Cost Variation Adjustments
- Labor items are percentage increases or decreases from 100% for labor units used in the book's labor unit columns. (See first page of each Division.)
- Material and Labor combined are percentage increases and decreases from 100% for metro variations used in the book's Unit Cost columns, Quick Estimating sections and Sq.Ft. Cost Section - see Page 3.
- Recommend Unit Costs of 10% less than Book Units for Residential Construction.

Metric Equivalents & U.S. Weights & Measures Modifiers - See Page 3A-18.

EXAMPLE:	UNIT	LABOR	MATERIAL	COST
2" x 4" Wood Studs & Plates - 8' @ 16" O.C.	BdFt	.60	.52	1.12
Add 35% Taxes & Insurance on Labor		.21	-	.21
Add 5% Material Sales Tax		-	.03	.03
				1.36
Add 10% General Conditions & Equipment				.14
				1.50
Add 10% Overhead & Profit				.15
BdFt Cost - Bid Type Estimating			BdFt	1.65
or SqFt Wall Cost w/ 3 Plates - Quick Estimating (page 6A-14)			or SqFt	1.40

See Division 4 for Other Examples (Masonry)

Purchasing Practice (usual) in the industry for each section of work is indicated immediately after section heading in parenthesis and defined as follows:

- L&M - Labor and Material Subcontractor - Firm bid to G.C. according to plans and specifications for all labor and material in that section or subsection.
- M - Material Subcontractor - Firm bid to G.C. according to plans and specifications for all material in that section or subsection.
- L - Labor Subcontractor - Firm bid to G.C. to *install only* material furnished according to plans and specifications by others.
- S - Material Supplier - Firm Price for materials delivered only.
- E - Equipment Supplier - Firm Price for Equipment Rented or Purchased.

Installation Practice by particular trades are listed behind each section heading and are according to "Agreement and Decisions Rendered Affecting the Building Industry," by the A.F.L. Building and Construction Trades Department, A.F.L.-C.I.O. Area variations, jurisdictional problems and multiple jurisdictions are possible.

DIVISION #1 - GENERAL REQUIREMENTS

		PAGE
0101.0	**GENERAL CONDITIONS - Specified (A.I.A. Doc. A201, 1987)**	1-2
.1	SUPERINTENDENT	1-2
.2	PERMITS AND FEES	1-2
.3	TAXES	1-2
.31	Sales and Use Tax	1-2
.32	Unemployment Compensation - State	1-2
.33	Unemployment Compensation	1-2
.34	F.I.C.A. - Employer's Share	1-2
.4	INSURANCE - CONTRACTOR'S LIABILITY	1-2
.5	INSURANCE - HOLD HARMLESS	1-2
.6	INSURANCE - ALL RISK	1-2
.7	CLEANUP	1-2
.8	PERFORMANCE AND PAYMENT BONDS	1-2
.9	SAFETY OF PERSONS AND PROPERTY	1-2
.10	TESTS AND SAMPLES	1-2
0102.0	**SPECIAL CONDITIONS (Changes to G.C. 0101 and Temporary Utilities and Facilities Specified)**	1-3
.1	TEMPORARY OFFICE	1-3
.2	TEMPORARY STORAGE SHED	1-3
.3	TEMPORARY SANITARY FACILITIES	1-3
.4	TEMPORARY TELEPHONE	1-3
.5	TEMPORARY POWER AND LIGHT	1-3
.6	TEMPORARY WATER	1-3
.7	TEMPORARY HEAT	1-3
.8	TEMPORARY ENCLOSURES AND COVERS	1-3
.9	TEMPORARY ROADS AND RAMPS	1-3
.10	TEMPORARY DEWATERING	1-3
.11	TEMPORARY SIGNS	1-3
.12	TEMPORARY FENCING, BARRICADES AND WARNING LIGHTS	1-3
0103.0	**MISCELLANEOUS CONDITIONS - Non-Specified**	1-4
.1	FIELD ENGINEERING AND LAYOUT	1-4
.2	TIMEKEEPERS AND ACCOUNTANTS	1-4
.3	SECURITY	1-4
.4	OPERATING ENGINEERS FOR HOISTING	1-4
.5	TRAVEL AND SUBSISTENCE - Field Engineers	1-4
.6	TRAVEL AND SUBSISTENCE - Office Employees	1-4
.7	OFFICE SUPPLIES AND EQUIPMENT	1-4
.8	COMMUNICATIONS EQUIPMENT	1-4
.9	HOISTING EQUIPMENT	1-4
.91	Towers and Hoists	1-4
.92	Cranes - Climbing - 20' Override	1-4
.93	Cranes - Mobile Lifting (No Operator)	1-4
.94	Fork Lifts	1-5
.95	Platform Lifts	1-5
.10	TRUCKS AND OTHER JOB VEHICLES (Site Use)	1-5
.11	MAJOR EQUIPMENT	1-5
.12	SMALL TOOLS	1-5
.13	HAULING EQUIPMENT AND TOOLS	1-5
.14	GAS AND OIL FOR EQUIPMENT	1-5
.15	REPAIRS TO EQUIPMENT AND TOOLS	1-5
.16	PHOTOGRAPHS - 2 Views, 4 Copies	1-5
.17	SCHEDULING - C.P.M., Pert, Network, Analysis	1-5
.18	COMPUTER COSTS	1-5
.19	PERFORMANCE BOND - SUBCONTRACTORS	1-5
.20	PUNCH LIST AND FINAL OUT	1-5
1A	**EQUIPMENT**	1-6A and 1-7A

DIVISION #1 - GENERAL REQUIREMENTS

General Requirements amount to approximately 6% to 15% of total job costs for General Construction and 5% to 10% for Residential Construction. All items below are included in G.R. except 0101.32, .33, and .34 (taxes and insurance on labor). See Square Foot Costs (SF Section) for more accurate Dollar Value of Job Costs.

			UNIT	LABOR	MATERIAL
0101.0	**GENERAL CONDITIONS - Specified (AIA Doc A201 1987)**				
.1	SUPERINTENDENT (Art. 3.9) Job Size/Complexity		Week	$1,600	-
.2	PERMITS AND FEES (Art. 3.7)				
	Concrete and Commercial	- Approx. Average	MCuFt	-	$2.50
	Residential (Finish Area)	- Approx. Average	SqFt	-	$ .50
	Remodeling	- Approx. Average	M$	-	$5.00
	Varies with Each Permit Division				
.3	TAXES (Art. 3.6)				
.31	Sales and Use Tax (Varies as to State)		%		Avg. 5.5%

		Payroll Max	Per $100 or % Gross Payroll	
.32	Unemployment Compensation	to $ 21,000	State Maximum	9.1%
	(Varies as to State)		State Minimum	.7%
.33	Unemployment Compensation	to $ 7,000	Fed. Maximum	.8%
.34	FICA - Employer's Share	to $ 84,900		6.20%
	Medicare	No Limit		1.45%
.4	INSURANCE - CONTRACTOR'S LIABILITY			
	(Worker's Compensation, P.L. & P.D.) (Art. 11.1.1)			
	Classification Avg - Contractors Experience		Gross Payroll	14.0%
	Clerical - Office	(#8810)	" "	1.0%
	Supervision	(#5606)	" "	3.0%
	Excavation	(#6217)	" "	12.0%
	Concrete - Structural - N.O.C.	(#5213)	" "	15.0%
	Slabs on Ground, Flat	(#5221)	" "	14.0%
	Masonry	(#5022)	" "	11.0%
	Iron and Steel - N.O.C.	(#5057)	" "	32.0%
	Carpentry - N.O.C. Commercial	(#5403)	" "	32.0%
	Interior & Dwellings	(#5437)	" "	16.0%
	Wallboard	(#5445)	" "	13.0%
	Plumbing			7.0%
	Electrical			6.0%
	Wrecking (Demolition)			35.0%
	PL. and PD.			1.5%
	Add for Assigned Risk			75%

			UNIT	LABOR	MATERIAL
.5	INSURANCE - HOLD HARMLESS				
	(By Owner) (Art. 11.2) (Contract Amount)		M$		$ .80
.6	INSURANCE - ALL RISK (Fire/Ext.Cov./Theft/etc.)				
	(By Owner) (Art. 11.2) ($1,000 Deductible)				
	Fire Proof - All Concrete		Annual)		$2.00
	Fire Resistive - Covered Steel		Insurable)		$2.50
	Ordinary - Masonry, Ext. Walls & Joists		Value)		$3.50
	Frame, Wood		per M$)		$6.00
	Deduct for $2,500 Deductible				10%
	Add for Light Protection Areas - Approx.				50%
.7	CLEANUP (Art. 3.15)				
.71	Progress		Bldg - SqFt	$.15	$ .05
.72	Glass, 2 Sides		Glass Area - SqFt	$.30	-
.8	PERFORMANCE & PAYMENT BOND		(If Specified) (Art. 11.4)		
	$ -0- to $ 99,999		Job Bureau Comp. Contr. Amount		2.50%
	$ 100,000 to $ 499,999				1.50%
	$ 500,000 to $2,499,999				1.00%
	$2,500,000 to $4,999,999				.75%
	$5,000,000 to $7,499,999				.70%
	$7,500,000 Up				.65%
.9	SAFETY OF PERSONS & PROPERTY (Art. 10.1 and 10.2) - Job Condition				
.10	TESTS AND SAMPLES (Art. 3.1) - See Appropriate Division				

See 0102.0 Temporary Fences and Barricades

DIVISION #1 - GENERAL REQUIREMENTS

0102.0 SPECIAL CONDITIONS (Changes to G.C. 0101 and Temporary Utilities & Facilities, usually Specified)

		UNIT	LABOR	MATERIAL
.1	TEMPORARY OFFICE			
	Fixed - 10' x 16" Wood	Each	1,350	2,300
	Mobile (Special Constr.) 8' x 32'	Each	900	8,500
		Month	-	400
	Mobile (Special Constr.) 10' x 50'	Each	1,000	12,500
		Month	-	650
	Mobile (Special Constr.) 12' x 60'	Each	1,200	14,500
		Month	-	750
	Add for Delivery & Pickup	Each	-	400
.2	TEMPORARY STORAGE SHED -12' x 24'	Each	1,700	3,300
.3	TEMPORARY SANITARY FACILITIES - Chemical Toilet	Month	-	100
.4	TEMPORARY TELEPHONE (No Long Distance) - Per line	Month	-	85
	TEMPORARY TELEPHONE	Initial	-	250
	TEMPORARY TELEPHONE - Cellular	Month	-	125
	TEMPORARY FAX	Initial	-	1,100
.5	TEMPORARY POWER AND LIGHT			
	Initial Service 110V	Job	-	1,200
	220V	Job	-	1,800
	440V	Job	-	3,300
	Monthly Service and Material	Month	-	225
	Temporary Wiring and Lighting-Total Job	SqFt	-	.08
.6	TEMPORARY WATER -Initial Service	Initial	Job Condition	
	Monthly Service	Month	-	75.00
.7	TEMPORARY HEAT (after Enclosed)			
	See Divisions 3 & 4 for Temporary Heat, Concrete & Masonry			
	See Division 1A for Equipment Rental Rates			
	Heaters:			
	Steam & Unit - Heater Rent - SqFt Htd Area	Month	-	.11
	Piping - SqFt Htd Area	Month	-	.13
	Condensate - SqFt Htd Area	Month	-	.18
	Mobile Home Heater Rent - SqFt Htd Area	Month	-	.10
	Fuel SqFt Htd Area	Month	-	.25
	Tending 8 - Hour/Day	8 Hrs	-	215.00
.8	TEMPORARY ENCLOSURES AND COVERS			
	Per Door - Wood	Each	30.00	40.00
	Per Window - Polyethylene and Frame	Each	12.00	10.00
	Walls - Frame and Reinforced Poly - 1 use	SqFt	.20	.17
	2 uses	SqFt	.18	.13
	Frame and 1/2" Plywood - 1 use	SqFt	.55	.95
	2 uses	SqFt	.45	.60
	Floors- 2" Fiberglass & 7 mil Poly- 1 use	SqFt	.10	.40
.9	TEMPORARY ROADS & RAMPS - No Surfacing	SqYd	Job Condition	
.10	TEMPORARY DEWATERING - No Tending			
	Pumping - 2" Pump	Day	-	90.00
	3" Pump	Day	-	100.00
	4" Pump	Day	-	115.00
	Add for Tending			
	Well Points - LnFt Header	Month	-	100.00
	Add Each Month	Month	-	75.00
.11	TEMPORARY SIGNS - 4' x 8'	Each	100.00	430.00
.12	TEMP. FENCING, BARRICADES & WARNING LIGHTS			
	6' Chain Link - Rented (Erected and Removed)	LnFt	1.15	4.50
	Add for Barbed Wire	LnFt	.30	.80
	Add for Gates	LnFt	7.50	17.00
	Snow Fence - New	LnFt	.38	1.10
	Concrete Barriers (Rented - Month)	LnFt	-	3.50
	Flasher Warning Light (Rented - Month)	Each	-	19.00
	8' Barricade (Rented - Month)	Each	-	40.00
	12' Barricade (Rented - Month)	Each	-	50.00

DIVISION #1 - GENERAL REQUIREMENTS

0103.0	MISCELLANEOUS CONDITIONS (Non-Specified)	UNIT	LABOR*	MATERIAL
.1	FIELD ENGINEERING AND LAYOUT	Week		Job Cond.
.2	TIMEKEEPERS AND ACCOUNTANTS	Week		Job Cond.
.3	SECURITY	Week		Job Cond.
.4	OPERATING ENGINEERS FOR HOISTING	Week		Job Cond.
.5	TRAVEL AND SUBSISTENCE - Field Employees			Job Cond.
.6	TRAVEL AND SUBSISTENCE - Office Employees			Job Cond.
.7	OFFICE SUPPLIES AND EQUIPMENT	Week		.03%
.8	COMMUNICATIONS EQUIPMENT			
	Radio-Multi Channel Hand Unit w/ Charger & Battery	Each	-	1,300
	Base Unit - F.M. with Antenna and Mike	Each	-	1,430
.9	HOISTING EQUIPMENT (Towers, Cranes and Lifts)			
.91	Towers and Hoists - Portable Self-Erecting Style			
	Skip (Based on 100' Unit) -3000# to 5000# Capacity	Month	-	2,200
	2000# to 3000# Capacity	Month	-	2,000
	Add to Heights Above 100'	LnFtMo	-	20
	Towers - Fixed (based on 100') - 30' Override			
	Tubular Frame - Double Well - 5000# Capacity	Month	-	1,800
	Installation - Double Well - Subcont - Up	LnFt	60	-
	Double Well Down	LnFt	45	-
	Add on or Deduct from 100' - Month	LnFt	-	20
	Add for Concrete Bucket and Hopper	Month	-	400
	Add for Chicago Boom	Month	-	230
	Add for Gates, Signals, etc (Safety)	Job	700	1,000
	Add for Fence Enclosure (Safety)	Job	440	900
	Add for Platform Enclosure (Safety)	Job	330	500
	Hoist Engine, Gas Powered - 2 Drum	Month	-	1,800
	Add for Electric Powered	Month	-	200
	Towers and Hoists - Personnel 6000# Capacity			
	Based on 100' Unit - 8 1/2' Override	Month	-	3,500
	Add Above 100'	LnFt	-	35
	Installation - Up	LnFt	-	40
	Down	LnFt	-	38
.92	Cranes - Climbing - 20' Override			
	Based on 125' Bldg-130' Boom, 4000#	Month	-	6,000
	to 200' Boom, 7000#	Month	-	8,000
	230' Boom, 7000#	Month	-	10,200
	Add for LnFt Tower above 125'	Month	-	35
	Installation			
	Foundation	Each	-	10,000
	440-Volt Transformer Install	Each	-	2,500
	Installation - Up & Down	LnFt	-	110
	Up	Each	-	15,000
	Down	Each	-	15,000
	Avg - Up, Down & Haul - 125' Tower	Each	-	30,000
	Add per Lift - Jacking	LnFt	-	50
	Add per Floor Pattern	Each	-	1,000
.93	Cranes - Mobile Lifting (no Operator)	HOUR	DAY	MONTH
	15-Ton Hydraulic	95	500	3,400
	20-Ton	100	550	3,500
	25-Ton	105	600	5,000
	30-Ton	120	650	5,500
	45-Ton	135	7500	6,400
	75-Ton Cable	155	800	7,500
	90-Ton	180	950	8,500
	25-Ton	210	1,300	12,000
	Add for Local Travel & Setup Time - 4 Hours			

*Subcontract

DIVISION #1 - GENERAL REQUIREMENTS

0103.0 MISCELLANEOUS CONDITIONS, Cont'd...

		UNIT	LABOR	MATERIAL
.94	Fork Lifts - Construction			
	Two-Wheel Drive - Gasoline Powered			
	21' 2500# Capacity	Month		2,000
	36' 4500# Capacity	Month		2,500
	Four-Wheel Drive - 21' 5000# Capacity	Month		2,500
	36' 6000# Capacity	Month		2,700
	38' 7000# Capacity	Month		3,000
	42' 8000# Capacity	Month		3,800
.95	Platform Lifts - Telescoping Boom - Rolling	Month		3,300
	Scissor - Rolling	Month		1,300
.10	TRUCKS AND OTHER JOB VEHICLES (Site Use)			
	$ 500,000 Job	JobAvg		1,600
	$1,000,000 Job	JobAvg		6,000
	$5,000,000 Job	JobAvg		14,000
	1-Ton Pickup	Month		600
	2-Ton Flatbed	Month		850
	2 1/2-Ton Dump	Month		1,400
.11	MAJOR EQUIPMENT (See Div. 2,3,4,5,& 6 if Subbed)			
	(See 1-6A & 7A: Rental Rates, Purchase Prices)			
	$ 100,000 Job	JobAvg		3,500
	$ 500,000 Job	JobAvg		7,400
	$1,000,000 Job	JobAvg		14,000
	$5,000,000 Job	JobAvg		55,000
	As a % of Labor (No Hoisting Equipment)			5%
.12	SMALL TOOLS AND SUPPLIES			
	(See 1-6A & 7A: Rental Rates, Purchase Prices)			
	$ 100,000 Job	JobAvg		1,300
	$ 500,000 Job	JobAvg		3,600
	$1,000,000 Job	JobAvg		6,000
	$5,000,000 Job	JobAvg		30,000
	As a Percentage of Labor	JobAvg		2%
.13	HAULING EQUIPMENT AND TOOLS			
	$ 100,000 Job	JobAvg	850	600
	$ 500,000 Job	JobAvg	1,600	1,500
	$1,000,000 Job	JobAvg	3,000	3,000
	$5,000,000 Job	JobAvg	10,000	8,000
.14	GAS AND OIL FOR EQUIPMENT			
	$ 500,000 Job	JobAvg		900
	$1,000,000 Job	JobAvg		1,400
	$5,000,000 Job	JobAvg		6,000
	As a Average of Equipment Cost	JobAvg		10%
.15	REPAIRS TO EQUIPMENT AND TOOLS			
	$ 100,000 Job	JobAvg		600
	$ 500,000 Job	JobAvg		1,200
	$1,000,000 Job	JobAvg		2,000
	$5,000,000 Job	JobAvg		7,000
	As an Average of Equipment Cost			10%
.16	PHOTOGRAPHS - 2 Views, 4 Copies	Month		125
.17	SCHEDULING - C.P.M., Pert, Network, Analysis			
	$ 500,000 Job	JobAvg		4,500
	$1,000,000 Job	JobAvg		8,000
	$5,000,000 Job	JobAvg		14,000
	Add for Update, Monitor	Each		700
.18	COMPUTER COSTS	Employee/Wk		8
.19	PERFORMANCE BOND - SUBCONTRACTORS	ContrAmt		1.5%
.20	PUNCH LIST AND FINAL OUT	ContrAmt		.5%

DIVISION #1 - GENERAL REQUIREMENTS

TOOLS & EQUIPMENT

See 103.9 for Towers, Cranes and Lift Equipment

	Approximate Cost New	Fair Rental* Rate per Mo.
Div. 1 General Conditions		
Engineering Instruments - Transits	1,170	320
Levels 18"	950	240
Fans - 15,000 CFM - 36"	1,100	400
Generators - 3500 Watt - Portable	1,200	420
5000 Watt - Portable	1,550	675
Heaters, Temp - Oil 100M to 300 BTU	1,300	525
LP Gas 90M BTU	360	200
150M BTU	400	270
350M BTU	710	460
Laser - Rotating	5,700	840
Light Stands - Portable - 500 watt Quartz	670	250
Pumps- Submersible 1/2 HP Elect 2"	420	160
Centrifugal 2" 8,000 Gal	970	580
3" 17,000 Gal	1000	590
Diaphragm 2" - 3" x 3" - Pneumatic	1,400	440
3" - 3" x 3" - Gas	1,450	570
Submersible - 5 HP 3"	1,500	760
Hose - Discharge - 3" - 25'	100	105
Suction - 3" - 20'	130	135
Vacuum Cleaners - Wet Dry - 5 Gal	600	160
Div. 2 Excavation (Including Demolition)		
Air Tools- Breakers Light Duty - 35 to 40#	960	270
Heavy Duty - 60 to 90#	1,320	375
Rock Drill - 40#	1,780	390
Chipping Hammer - 7# to 12#	770	340
Clay Digger	720	300
Hose - 3/4" - 50'	130	65
Compressors 125 cfm Gas - Tow	10,700	770
185 cfm Diesel - Trailer Mounted	13,600	1,080
26' Conveyor - 2 HP Electric	9,150	1,175
Electric Breakers - 60#	2,100	450
Jacks - Hydraulic - 20 Ton - 11 HP	800	130
Screw - 20 Ton	360	60
Saws - Chain 18"	550	240
Concrete Asphalt 14" Blade	1,600	400
24" Plate - Vibratory Gas 170#	2,100	730
Tampers - Rammer Gas 135#	2,300	780
Div. 3 Concrete		
Bucket & Boot- 1 CuYd	2,050	400
1 1/2 CuYd	2,600	420
2 CuYd	3,400	440
Buggies - Manual - 6 to 8 CuFt	470	115
Power Operated - 16 CuFt	6,000	875
Conveyors - 32' Hydraulic and Belt - Gas	9,400	1,200
50' Hydraulic and Belt - Gas	20,800	2,500
Core Drill - Diamond - 15 Amp	3,250	620
with Trailer	9,400	1,025
Crack Chaser	4,200	1,430
Drill - Cordless 3/8" Reversing	290	95
Heavy Duty 1/2" Reversing	370	200
3/4" Reversing	670	300
Sander- Hand - 9"	330	200
Grinder- Ceiling 7' to 11' - Elec	2,200	260
Floor - Elec - 1 1/2 HP	3,050	590
Hammer - Drill - 3/4" Variable Speed	470	225
Roto Chipping - 1 1/2"	500	420
Hopper - Floor 30 CuFt	1,100	210
Mixer - 6 CuFt	2,300	510
Planer - 10'	4,300	720
Scarifier & Scabblers - Single Bit	1,100	390
5 Bit Floor - 175 cfm	7,800	2,900
Stripper - Floor	3,750	560

*Weekly Rent 40% of monthly, Daily Rent 15% of monthly, Hourly Rent 2% of monthly.

DIVISION #1 - GENERAL REQUIREMENTS
TOOLS & EQUIPMENT

		Approximate Cost New	Fair Rental* Rate per Mo.
Div. 3	**Concrete, Cont'd...**		
	Saws - Rotary Hand - 7 1/4"	190	90
	9 1/4" - 8 1/4"	235	115
	Table 14" Radial Arm 3 HP	1,800	300
	Floor - Gas 11 HP	1,550	480
	Gas 18 HP	4,300	690
	Electric 8 HP 14"	1,820	510
	Cut off 14" Electric	710	380
	Screeds - Vibrating 6'	900	220
	12 1/2' Beam Type	1,950	350
	25' Beam Type	4,200	950
	Troweling Machines - Gas 36" 4 Blade 7 HP	1,700	500
	48" 4 Blade 6 HP	1,950	540
	Trunks & Tremmies - 12" and Hopper	180	60
	Vibrators - Flexible Head - 1 HP	1,100	330
	2 HP	1,200	430
	High cycle - 180 Cycle	3,200	500
	Back Panel 2 1/2 HP	750	230
	Wheelbarrows	115	50
Div. 4	**Masonry**		
	Hoist Electric - One ton	1,050	170
	Mixer - Mortar - 6 Cu Ft - Gas 7 HP	2,000	430
	Elec 1 1/2 HP	2,120	400
	8 Cu Ft - Gas 7 HP	2,400	550
	12 Cu Ft - Gas or Electric	3,600	800
	Pallet Jack	740	140
	Pump Grout - Hydraulic 10 GPM	6,200	1,400
	Saws - Table - 1 1/2 HP - 14"	2,150	300
	5 HP - 14"	3,100	580
	Scaffold - 5' - 6" with Braces	75	3.60
	6' - 6" with Braces	80	4.00
	Wheels	85	11.00
	Brackets	22	2.00
	Plank - Aluminum	90	5.00
	16' Laminated	35	3.00
	Base - Adj	25	1.60
	Splitter - 10,000#	1,700	350
	Rolling Towers	350	90
	Swing Stage - Motorized w/ deck-cables-safety	9,000	1,100
Div. 6	**Carpentry**		
	Door Hanging Eq. (Router, Mortiser, Template)	300	100
	Drills - Heavy Duty (See Div. 3)		
	Fasteners - Power - Drywall	150	40
	Hammer - Drill 1/2"	310	120
	Miter Box - 14" Power	410	140
	Nailers and Staplers - Spot w/ compressor	1,800	260
	Planers	500	150
	Sanders - Disc 9"	260	90
	Belt 3" - Heavy Duty	300	115
	Floor	420	85
	Saws - Band	300	125
	Chain	350	210
	Rotary, Hand - 7 1/4"	190	90
	8 1/4"	220	120
	Table - Fixed 10"	400	140
	Jig	200	80
	Reciprocating	450	220
	Cut-off - 12" Elec.	2,000	450
	Scaffold - Rolling	450	110
	Screwdriver - Variable Speed - Drywall	250	110
	Stud Drivers - Low Velocity	280	120
	High Velocity	400	140
	Wrench - 1/2" Impact - Air	290	140
	1" Impact - Air	600	250

*Weekly Rent 40% of monthly, Daily Rent 15% of monthly, Hourly Rent 2% of monthly.

DIVISION #1 - GENERAL REQUIREMENTS

DIVISION #2 - SITEWORK & DEMOLITION

		PAGE
0201.0	**DEMOLITION & CLEARING (CSI 02100)**	**2-3**
.1	STRUCTURE MOVING	2-3
.2	CLEARING & GRUBBING	2-3
.3	DEMOLITION	2-3
.31	Total Building	2-3
.32	Selective Building	2-3
.33	Site	2-4
.34	Disposal Haul Away	2-4
.35	Cutting, Core Driller and Blasting	2-4
0202.0	**EARTHWORK (CSI 02200)**	**2-5**
.1	GRADING	2-5
.2	EXCAVATION	2-5
.3	BACKFILL	2-5
.4	COMPACTION	2-5
.5	DISPOSAL	2-6
.6	TESTS AND SUBSURFACE EXPLORATION	2-6
0203.0	**PILE AND CAISSON FOUNDATIONS (CSI 02300)**	**2-6**
.1	CONCRETE PILING	2-6
.11	Augered and Cast-In-Place Concrete	2-6
.12	Precast Prestressed	2-6
.2	STEEL	2-6
.21	Steel Pipe	2-6
.22	Steel H. Section	2-6
.3	WOOD	2-6
.4	CAISSONS	2-7
0204.0	**EARTH AND WATER RETAINER WORK (CSI 02400)**	**2-7**
.1	STREET SHEET PILING	2-7
.2	WOOD SHEET PILING AND BRACING	2-7
.3	H. PILES AND WOOD SHEATHING	2-7
.4	UNDERPINNING	2-7
.5	RIP RAP AND LOOSE STONEWALLS	2-7
.6	CRIBBING	2-7
0205.0	**SITE DRAINAGE (CSI 02500)**	**2-7**
	(See Divisions 15 and 16 for Site Utilities)	
.1	DITCHING AND BACKFILL (OPEN)	2-7
.2	DEWATERING	2-7
.3	CULVERTS	2-8
.4	CATCH BASINS AND MANHOLES	2-8
.5	FOUNDATION DRAINAGE	2-8
0206.0	**PAVEMENTS, CURBS AND WALKS (CSI 02600)**	**2-9**
.1	PAVING	2-9
.11	Bituminous	2-9
.12	Concrete	2-9
.13	Stabilized Aggregate	2-9
.14	Brick	2-9
.2	CURBS AND GUTTERS	2-10
.21	Concrete - Cast in Place	2-10
.22	Concrete - Precast	2-10
.23	Bituminous	2-10
.24	Granite	2-10
.25	Timbers, Treated	2-10
.26	Plastic	2-10

DIVISION #2 - SITEWORK & DEMOLITION

0206.0		**PAVEMENTS AND WALKS, CONT'D...**	
	.3	WALKS	2-10
	.31	Bituminous	2-10
	.32	Concrete	2-10
	.33	Precast Block	2-10
	.34	Crushed Rock	2-10
	.35	Flagstone	2-10
	.36	Brick	2-10
	.37	Wood	2-10
0207.0		**FENCING**	**2-11**
	.1	CHAIN LINK	2-11
	.2	WELDED WIRE	2-11
	.3	MESH WIRE	2-11
	.4	WOOD	2-11
	.5	GUARD RAILS	2-11
0208.0		**RECREATIONAL FACILITIES**	**2-12**
	.1	PLAYING FIELDS AND COURTS	2-12
	.2	RECREATIONAL EQUIPMENT	2-12
	.3	SITE FURNISHINGS	2-12
	.4	SHELTERS	2-12
0209.0		**LANDSCAPING (CSI 02900)**	**2-13**
	.1	SEEDING	2-13
	.2	SODDING	2-13
	.3	DISCING	2-13
	.4	FERTILIZING	2-13
	.5	TOP SOIL	2-13
	.6	TREES AND SHRUBS	2-13
	.7	BEDS	2-13
	.8	WOOD CURBS AND WALLS	2-13
0210.0		**SOIL STABILIZATION**	**2-14**
	.1	CHEMICAL INTRUSIONS	2-14
	.2	CONCRETE INTRUSIONS	2-14
	.3	VIBRO-FLOTATION	2-14
0211.0		**SOIL TREATMENT**	**2-14**
	.1	PESTICIDES	2-14
	.2	VEGETATION	2-14
0212.0		**RAILROAD WORK (CSI 02450)**	**2-14**
	.1	NEW WORK	2-14
	.2	REPAIR WORK	2-14
0213.0		**MARINE WORK (CSI 02480)**	**2-14**
0214.0		**TUNNEL WORK (CSI 02300)**	**2-14**

See Division 1501 for Piped Utility Material (CSI 02600)
See Division 1501 for Water Distribution (CSI 02660)
See Division 1501 for Fuel Distribution (CSI 02680)
See Division 1503 for Sewage Distribution (CSI 02700)
See Division 1600 for Power Communication (CSI 02780)

DIVISION #2 - SITEWORK & DEMOLITION

0201.0 DEMOLITION & CLEARING (Op.Eng., Truck Dr. & Lab.)

	UNIT	COST with MACHINE
.1 STRUCTURE MOVING - Wood Frame	SqFt	13.50
With Masonry	SqFt	16.50
.2 CLEARING AND GRUBBING		
Trees - Trunk	Dia/Inch	13.50
Stump - Removed	Dia/Inch	10.50
Chipped - Below Grade	Dia/Inch	6.00
Add for Hauling Away and Dumping	Dia/Inch	5.00
Shrubs & Brush - Light	Acre	1,800.00
Medium	Acre	2,600.00
Heavy	Acre	3,400.00

.3 DEMOLITION AND LOAD (See .34 for Disposal)
.31 Total Building

	UNIT	COST
Wood - Mainly Housing	CuFt/Bldg	.28
Masonry - Load Bearing	CuFt/Bldg	.34
Concrete - Frame	CuFt/Bldg	.42
Steel - Frame	CuFt/Bldg	.34
Add for Floor Above 2 Stories	CuFt/Bldg	.06

.32 Selective Building Removals (No Cutting/Disposal)
    (See .35 - Cut and Drill)
    (See .34 - Disposal/Haul Away)
    (See .33 - Site Removal)

	Unit	Labor Only	Machine Only	Unit	Labor Only	Machine Only
Concrete:						
8" Walls Reinforced	CuYd	235.00	90.00	SqFt	6.00	1.75
Non-Reinforced	CuYd	185.00	80.00	SqFt	4.60	1.50
12" Walls Reinforced	CuYd	320.00	70.00	SqFt	11.60	2.60
Footings - 24" x 12"	CuYd	255.00	155.00	LnFt	9.50	5.70
6" Struc. Slab Reinforced	CuYd	200.00	62.00	SqFt	3.75	1.12
8" Struc. Slab Reinforced	CuYd	212.00	83.00	SqFt	5.35	1.53
4" Slab on Ground Reinforced	CuYd	380.00	52.00	SqFt	1.70	.61
Non-Reinforced	CuYd	340.00	45.00	SqFt	1.30	.57
6" Slab on Ground Reinforced	CuYd	215.00	42.00	SqFt	2.10	.77
Non-Reinforced.	CuYd	205.00	39.00	SqFt	1.55	.73
9" Stairs Reinforced	CuYd	270.00	90.00	SqFt	6.40	2.50
Masonry:						
4" Brick or Stone Walls	CuYd	99.00	32.00	SqFt	1.20	.38
and 8" Backup	CuYd	180.00	68.00	SqFt	2.20	.68
Block or Tile Partitions	CuYd	77.00	30.00	SqFt	1.00	.34
6" Block or Tile Partitions	CuYd	62.00	23.00	SqFt	1.12	.42
8" Block or Tile Partitions	CuYd	54.00	42.00	SqFt	1.32	.50
12" Block or Tile Partitions	CuYd	145.00	57.00	SqFt	1.70	.62
Add for Plastered Type				SqFt	.38	.11

Misc - Hand Work (Add for Equipment & Scaffold)

	UNIT	Labor	
Acoustical Ceilings - Attached (Incl. Iron)	SqFt	.54	-
Suspended (Incl. Grid)	SqFt	.34	-
Asbestos - Ceilings and Walls	SqFt	13.80	-
Columns and Beams	SqFt	34.00	-
Pipe	LnFt	42.50	-
Tile Flooring	SqFt	1.60	-
Cabinets and Tops	LnFt	9.60	-
Carpet	SqFt	.29	-
Ceramic and Quarry Tile	SqFt	1.10	-
Doors and Frames - Metal	Each	48.00	-
Wood	Each	43.00	-
Drywall Ceilings - Attached (Including Grid)	SqFt	.69	-
Wood or Metal Studs - 2 Sides	SqFt	.74	-
Paint Removal - Doors and Windows	SqFt	.69	-
Walls	SqFt	.54	-
Plaster on Wood or Metal Studs	SqFt	1.00	-
Ceilings - Attached (Incl. Iron)	SqFt	.95	-
Roofing - Built-Up	SqFt	.90	-
Shingles - Asphalt and Wood	SqFt	.30	-
Terrazzo Flooring	SqFt	1.60	-
Vinyl Composition Flooring (No Mastic Removal)	SqFt	.32	-
Wall Coverings	SqFt	.59	-
Windows - Metal or Wood	Each	32.00	-
Wood Flooring	SqFt	.34	-

DIVISION #2 - SITEWORK & DEMOLITION

0201.0 DEMOLITION & CLEARING, Cont'd...

.33 Site Removals
(See .34 for Disposal)
(See .35 for Cutting Additions)

	UNIT	LABOR ONLY (1)	MACHINE ONLY
Concrete: 4" Sidewalks - Reinforced	SqFt	1.80	.62
Non-Reinforced	SqFt	1.42	.59
6" Drives - Reinforced	SqFt	2.15	.80
Non-Reinforced	SqFt	1.67	.84
Curb - 6" x 18"	LnFt	5.00	1.50
and Gutter	LnFt	8.00	2.00
Asphalt: 2"	SqFt	1.10	.38
Curb	LnFt	2.55	.68
Fencing: 8" Metal	LnFt	1.65	.42

.34 Disposal or Haul Away (Including Truck or Box & Driver)

		COST
Haul - No Dump Charges Included		
Concrete Truck Measure - 75¢ a minute or	CuYd/Mile	.70
Masonry - 75¢ a minute or	CuYd/Mile	.67
Wood - 75¢ a minute or	CuYd/Mile	.62
Box Rent (Included Above) - 30 Cu.Yd.	Box	360.00
Clean Material - 20 Cu.Yd.	Box	325.00
10 Cu.Yd.	Box	275.00
Dump Charges - Concrete	CuYd	6.60
Building Materials	CuYd	9.60
Tree and Brush Removal	CuYd	13.20

.35 Saw Cutting/Core Drilling/Torch Cutting/Blasting

Saw Cutting (includes Operator):			
Slabs - Asphalt	LnFt	per inch	1.12
Concrete - Structural	LnFt	per inch	2.05
Hollow	LnFt	per inch	.85
Slabs on Grd/Non - Rein.	LnFt	per inch	.75
/with Mesh	LnFt	per inch	.95
Walls - Concrete - Reinforced	LnFt	per inch	3.20
Non - Reinforced	LnFt	per inch	2.60
Masonry - Hollow	LnFt	per inch	1.85
Solid	LnFt	per inch	2.40
Core Drilling:			
Solid Slab - 2" Diameter	Inch		3.65
3" Diameter	Inch		4.00
4" Diameter	Inch		5.50
6" Diameter	Inch		8.00
8" Diameter	Inch		10.80
Hollow Core Slab - 2" Diameter	Inch		2.35
3" Diameter	Inch		2.55
4" Diameter	Inch		3.70
6" Diameter	Inch		6.15
Torch Cutting - Steel per Inch	LnFt		5.10
Blasting - Unrestricted	CuYd		110.00
Restricted	CuYd		195.00
Cut Openings:			
Concrete (see .32)	SqFt		-
Sheet Rock Partitions	SqFt		3.00
Wood Floors	SqFt		2.40
No Disposal or Haul Away Included (see .34)			

---

(1) With Pneumatic Tools

DIVISION #2 - SITEWORK & DEMOLITION

0202.0   EARTHWORK (Op. Eng., Truck Driver & Lab.)

.1   GRADING

Machine	UNIT(1)	COST	UNIT(1)	COST
Strip Top Soil - 4"	CuYd	4.95	SqYd	.55
Spread Top Soil				
4" Site Borrow	CuYd	5.40	SqYd	.60
Off-Site Borrow ($13 CuYd)	CuYd	20.70	SqYd	2.30

Rough Grading - Cut and Fill	UNIT	SOFT(1)	MED.(1)	HARD(1)
Dozer - To 200'	CuYd	3.40	3.65	4.10
To 500'	CuYd	3.80	5.40	6.70
Scraper - Self Propelled - To 500'	CuYd	3.05	3.40	4.00
To 1,000'	CuYd	3.85	4.50	5.30
Grader - To 200'	CuYd	3.35	3.90	5.60
Add for Compaction (See 0202.4)				
Add for Hauling Away (See 0202.5)				
Add for Truck Haul on Site	CuYd	2.70	-	-
Add for Borrow Brought In				
Loose Fill – 3.00 Ton	CuYd	10.70		
Crushed Stone - 5.50 Ton	CuYd	14.00	-	-
Gravel - 4.30 Ton	CuYd	12.30		

Hand and Machine	UNIT	HAND LABOR	MACHINE
Fine Grading - 4" Fill - Site	CuYd	11.35	5.70
or SqFt		.14	.07
4" Fill - Building	CuYd	20.00	9.80
or SqFt		.25	.12

.2   EXCAVATION - Dig & Cast or Load (No Hauling)

Soil			
Open - Soft (Sand) w/Backhoe or Shovel	CuYd	20.5	3.30
Medium (Clay)	CuYd	28.50	4.00
Hard	CuYd	43.60	5.60
Add for Clamshell or Dragline	CuYd	-	1.30
Deduct for Front End Loader	CuYd	-	1.10
Deduct for Dozer	CuYd	-	.27
Trench or Pocket - Soft w/Backhoe or Shovel	CuYd	22.70	4.80
Medium	CuYd	28.90	6.20
Hard	CuYd	41.50	7.50
Add for Clamshell or Dragline	CuYd	-	1.20
Augered - 12" diameter	CuYd	47.80	9.00
24" diameter	CuYd	36.50	6.10
Add for Frost	CuYd	-	7.20
Rock - Soft - Dozer	CuYd	-	32.00
Hammer	CuYd	-	78.00
Medium - Blast	CuYd	-	95.00
Hammer	CuYd	-	160.00
Hard - Blast	CuYd	-	210.00
Hammer	CuYd	-	280.00
Swamp	CuYd	-	9.00
Underwater	CuYd	-	11.80

.3   BACKFILL (Not Compacted and Site Borrow)   CuYd   10.15   2.60
   Add for Off-Site Borrow   CuYd   -   10.50

Add to <u>All Above</u> for Mobilization Average   5%

---
(1) Machine and Operators

DIVISION #2 - SITEWORK & DEMOLITION

0202.0 EARTHWORK, Cont'd...

	UNIT	(1) LABOR	(2) COST
.4 COMPACTION (Incl. Backfilling & Site Borrow)			
Labor Only			
Building to 85 - 24" Lifts	CuYd	15.00	-
to 95 - 12" Lifts	CuYd	17.50	-
to 100 - 6" Lifts	CuYd	20.80	-
Labor Compaction & Machine Backfill Combined			
Building to 85 - 24" Lifts	CuYd	9.00	2.80
to 90 - 18" Lifts	CuYd	9.50	2.90
to 95 - 12" Lifts	CuYd	10.00	3.00
to 100 - 6" Lifts	CuYd	11.00	3.40
Subgrade (Including Grading and Rollers)	CuYd	-	3.80
Machine Only (No Grading)			
Site to 85 - 24" Lifts	CuYd		3.20
to 90 - 18" Lifts	CuYd		3.30
to 95 - 12" Lifts	CuYd		3.80
to 100 - 6" Lifts	CuYd		4.50
Add to Above for Off-site Borrow	CuYd		10.50
.5 DISPOSAL (Haul Away) Country	CuYd/Mile		.66
City	CuYd/Mile		.77
.6 TESTS			
Field Density - Compaction	Each		36.00
Soil Gradation	Each		52.00
Density (Proctor)	Each		84.00
Moisture and Specific Gravities	Each		90.00
Hydrometer (Sand, Silt and Clay)	Each		100.00
Rock -2" Core	LnFt		65.00
4" Core	LnFt		105.00
Add for Soil Penetration	LnFt		7.25
Add for Mileage	Mile		.42

0203.0 PILE AND CAISSON FOUNDATIONS (L&M)
(Operating Engineers and Iron Workers)

	UNIT	COST
.1 CONCRETE		
.11 Augured or Cast in Place -12"	LnFt	20.50
.12 Precast Prestressed -10"	LnFt	18.00
12"	LnFt	21.00
14"	LnFt	23.00
16"	LnFt	28.00
.13 Pipe Casing Pulled - Cast - in - Place - 10"	LnFt	19.00
12"	LnFt	20.00
.2 STEEL		
.21 Steel Pipe -10"	LnFt	24.00
12"	LnFt	28.00
14"	LnFt	32.00
16"	LnFt	35.00
Concrete - Filled - 10"	LnFt	27.00
12"	LnFt	33.00
14"	LnFt	39.00
16"	LnFt	44.00
.22 Steel H Section 8" - 36#	LnFt	25.00
10" - 42#	LnFt	28.00
12" - 53#	LnFt	34.00
14" - 73#	LnFt	36.00
14" - 89#	LnFt	43.00
.3 WOOD (Treated) to 40' (Incl. Tests & Cutoffs)		
10"	LnFt	17.00
12"	LnFt	18.00
14"	LnFt	20.00
16"	LnFt	21.00
Add for Mobilization		10%

---

(1) Hand Work
(2) Machine and Operators

DIVISION #2 - SITEWORK & DEMOLITION

		UNIT	COST
0203.0	**PILE & CAISSON, Cont'd...**		
.4	CAISSONS -20" Concrete Filled	LnFt	35.00
	24"	LnFt	42.00
	30"	LnFt	56.00
	36"	LnFt	72.00
	42"	LnFt	93.00
	48"	LnFt	118.00
	54"	LnFt	152.00
	60"	LnFt	190.00
	Add for Bells	Each	1,800.00
	Add for Wet Ground	LnFt	5.00
	Add for Obstructions - Hand Removal	CuFt	58.00
	Add for Obstructions - Machine Removal	CuFt	10.00
	Add for Reinforced Steel	Ton	1,550.0
	Add for Mobilization		10%
0204.0	**EARTH & WATER RETAINER WORK**		
.1	STEEL SHAFT PILING (Including Bracing) (L&M)		
	Heavy -27# -12"-16' Full Salvage	SqFt	17.00
	40# -15"-16' Full Salvage	SqFt	18.00
	57# -18"-16' Full Salvage	SqFt	19.00
	Add per Foot Longer than 16'	SqFt	2.00
	Add for No Salvage	SqFt	8.00
	Light -12 ga - 18" Black to 8' Full Salvage	SqFt	14.00
	10 ga - 18" Black to 8' Full Salvage	SqFt	15.00
	8 ga - 18" Black to 8' Full Salvage	SqFt	19.00
	Add for No Salvage	SqFt	7.00
	Add for Galvanized	SqFt	30%
.2	WOOD SHEET PILING & BRACING		
	6' Deep	SqFt	5.25
	8' Deep	SqFt	5.50
	10' Deep	SqFt	6.00
	12' Deep	SqFt	6.50
.3	H PILES & WOOD SHEATHING (4" Lbr.)		
	10' Deep	SqFt	20.00
	15' Deep	SqFt	21.00
	20' Deep	SqFt	26.00
	25' Deep	SqFt	30.00
	Deduct for Salvage	SqFt	40%
.4	UNDERPINNING	CuYd	670.00
.5	RIP RAP & LOOSE STONE WALLS		
	Machine Placed	CuYd	16.00
	Hand Placed	CuYd	51.00
	and Grouted	CuYd	120.00
.6	CRIBBING		
	6" Open to 10'	SqFt	22.00
	8" Open	SqFt	26.00
0205.0	**SITE DRAINAGE (Including Building Foundations)**		
.1	DITCHING AND BACKFILL (Open) - See 0202.2		
.2	DEWATERING		
	Pumping -2" Pump	Day	55.00
	3" Pump	Day	84.00
	4" Pump	Day	112.00
	Well Points - Ln.Ft. Header - first month	Month	125.00
	Add for Each Additional Month	Month	65.00
	Add for Attending Time	Hour	44.00

DIVISION #2 - SITEWORK & DEMOLITION

0205.0 SITE DRAINAGE, Cont'd...

	UNIT	LABOR	MATERIAL
.3 CULVERTS & PIPING (No Excavation - See 0202.2)			
.31 Corrugated Galvanized Metal			
Watertight (Bituminous Coated)			
12" Diameter w/Bands & Gaskets	LnFt	4.00	12.00
15"	LnFt	4.30	16.00
18"	LnFt	4.75	18.00
24"	LnFt	5.65	22.00
30"	LnFt	6.70	29.00
36"	LnFt	8.00	30.00
48"	LnFt	17.00	46.00
Non - Watertight			
12" Diameter w/Bands & Gaskets	LnFt	3.20	6.00
15"	LnFt	3.80	7.00
18"	LnFt	4.60	9.00
24"	LnFt	6.40	11.00
30"	LnFt	8.00	13.00
36"	LnFt	9.20	23.00
48"	LnFt	10.00	32.00
Add for Oval Arch Shape	LnFt	20%	-
Aprons			
18"	Each	16.00	60.00
24"	Each	18.00	90.00
30"	Each	20.00	150.00
36"	Each	23.00	240.00
.32 Reinforced Concrete Pipe - #3 - Class 5			
12" Diameter	LnFt	5.30	11.00
18"	LnFt	7.20	18.00
24"	LnFt	10.00	24.00
27"	LnFt	12.10	36.00
30"	LnFt	15.25	45.00
36"	LnFt	18.50	50.00
48"	LnFt	27.50	70.00
Add for Each Added Foot in Depth	LnFt	5%	-
Add for Apron 36" (Flared End)	Each	155.00	460.00
Add for Trash Guards (Galvanized)	Each	135.00	800.00
.33 Vitrified Clay 12"	LnFt	5.50	11.50
24"	LnFt	8.80	40.00
.4 CATCH BASINS & MANHOLES (No Excavation)			
48" Cast in Place (Concrete)	LnFt	140.00	145.00
48" Brick or Block	LnFt	135.00	150.00
48" Precast Concrete	LnFt	90.00	86.00
48" Precast Concrete Collar	LnFt	115.00	150.00
Add for Cover - Precast Concrete	Each	62.00	135.00
Add for Base - Precast Concrete	Each	83.00	140.00
Add for Covers & Grates - H.D.C.I.	Each	37.00	160.00
Add for Covers & Grates - L.D.C.I.	Each	37.00	160.00
Add for Covers - H.D. Watertight	Each	37.00	275.00
Add for Steps	Each	-	14.00
Add for Adjusting Rings 2"	Each	13.50	14.00
.5 FOUNDATION DRAINAGE (Subdrainage)			
4" Clay Pipe	LnFt	1.10	1.55
4" Corrugated - Perforated	LnFt	.90	.95
4" Plastic Pipe - Perforated	LnFt	.90	1.15
Solid	LnFt	.90	.90
6" Clay Pipe	LnFt	1.30	1.75
6" Corrugated - Perforated	LnFt	1.05	1.35
6" Plastic Pipe - Perforated	LnFt	1.05	2.25
Solid	LnFt	1.00	1.80
Add for Porous Surround 2'x 2' ($8.00 CuYd)	LnFt	1.30	2.60

DIVISION #2 - SITEWORK & DEMOLITION

0206.0 PAVEMENT, CURBS AND WALKS

.1   PAVING (PARKING LOTS AND DRIVEWAYS)

	UNIT	COST
.11 Bituminous (Material $35/Ton, 5% Asphalt, 10-Mile Delivery)		
1 1/2" Wearing 10.5 SqYd/Ton	SqYd	4.60
2" Wearing 9.0 SqYd/Ton	SqYd	5.85
2 1/2" Wearing 7.5 SqYd/Ton	SqYd	6.70
3" Wearing 6.0 SqYd/Ton	SqYd	7.80
Add for Paths and Small Driveways	SqYd	4.30
Add for Patching	SqYd	9.80
Add for Asphalt Content per 1%	SqYd	2.00
Add for Over 10-mile Delivery - Ton/Mile	SqYd	.35
Deduct for Base Asphalt	SqYd	10%
Add for Seal Coat	SqYd	1.55
Add for Striping - Paint	LnFt	.25
Add for Striping - Plastic	LnFt	.50
Base Course for Above - Add:		
4" Sand & Gravel $8.80 Ton and 1/3 Ton	SqYd	4.10
4" Crushed Stone $10.90 Ton	SqYd	4.90
6" Sand & Gravel $7.70 Ton and 1/2 Ton	SqYd	5.00
6" Crushed Stone $10.00 Ton	SqYd	6.00
Add per Mile over 10 Miles	SqYd	.40
Deduct for Belly Dump Delivery - $.75 Ton	SqYd	.40
4" Sand & Gravel	Ton	9.00
4" Crushed Stone	Ton	10.00
6" Sand & Gravel	Ton	8.30
6" Crushed Stone	Ton	9.40
Add per Ton Mile Over 10 Miles	Ton	.40
Deduct for Belly Dump Delivery	Ton	.90
.12 Concrete (4000# Ready Mix, $74 CuYd - Machine Placed)		
6" Reinforced with 6 x 6, 8-8 Mesh	SqYd	22.00
7" Reinforced with 6 x 6, 6-6 Mesh	SqYd	23.00
8" Reinforced with #4 Rods - 12" O.C.	SqYd	26.00
9" Reinforced with #5 Rods - 12" O.C.	SqYd	28.00
10" Reinforced with #6 Rods - 8" O.C.	SqYd	29.00
12" Reinforced with #8 Rods - 8" O.C.	SqYd	37.00
Add for Base - Same as Base Prices Above		
Add to Above for Hand Placed	SqYd	25%
Add for Driveways	SqYd	50%
.13 Stabilized Aggregate (Delivery by Dump Truck)		
4" Gravel - $8.80 Ton - 3/4" Compacted	SqYd	5.00
4" Crushed Stone - $10.00 Ton - 3/4" Compacted	SqYd	6.00
6" Gravel - $8.00 Ton - 3/4" Compacted	SqYd	5.80
6" Crushed Stone - $9.30 Ton - 3/4" Compacted	SqYd	7.50
6" Pulverized Concrete	SqYd	8.00
Add for Soil Cement Treatment	SqYd	.95
Add for Oil Penetration Treatment	SqYd	3.00
Add for Calcium Chloride Treatment	SqYd	.50
Add per CuYd Mile Over 10 Miles	SqYd	.40
Deduct for Belly Dump Truck Delivery	SqYd	.40
.14 Brick - 4" x 8" x 2 1/2"	SqYd	75.00
Add for Sand Bed - Compacted	SqYd	6.50
Add for Mortar Setting - Bed & Joints	SqYd	7.00

DIVISION #2 - SITEWORK & DEMOLITION

0206.0 PAVEMENT, CURBS AND WALKS, Cont'd...

		UNIT	LABOR	MATERIAL
.2	**CURBS & GUTTERS** (Cost Incl. Excavation & Equip.)			
.21	Concrete - Cast in Place			
	Curb			
	6" x 12" Small Job - 200' Day	LnFt	5.40	2.80
	6" x 18"	LnFt	6.50	3.30
	6" x 24"	LnFt	8.20	4.30
	6" x 30"	LnFt	9.50	5.10
	Add for Reinforced - Two #5 Rods	LnFt	.58	.90
	Add for Gutter - 6" x 12"	LnFt	4.00	2.20
	Add for Gutter - 6" x 18"	LnFt	4.25	3.05
	Add for Large Job - Over 200' Day	LnFt		10%
	Add for Formless 1000' Day Up	LnFt		20%
	Add for Curved or Radius Work	LnFt		40%
	Curb & Gutter Rolled			
	6" x 12" Small Job	LnFt	6.80	4.15
	6" x 18"	LnFt	8.00	5.60
	6" x 24"	LnFt	9.60	7.25
	Deduct for Large Job - Over 200' Day	LnFt	-	10%
	Deduct for Formless 1000' Day Up	LnFt	-	25%
	Add for Reinforced - Two #5 Rods	LnFt	.63	.80
	Add for Curved or Radius Work	LnFt	-	40%
	Add for Approaches - 8' x 16' x 6'	SqFt	3.40	2.10
.22	Concrete - Precast & Pinned - 6" x 6'	LnFt	1.80	6.95
	6" x 7'	LnFt	2.00	6.45
	6" x 8'	LnFt	1.90	5.90
.23	Asphalt - 6" x 8'	LnFt	.90	1.30
	Add for Curved Work	LnFt	-	30%
.24	Granite - 6" x 16"	LnFt	5.10	23.00
.25	Timbers, Treated - 6" x 6"	LnFt	1.80	3.40
	6" x 8"	LnFt	2.30	5.00
.26	Plastic - 6" x 6"	LnFt	1.60	4.90
.3	**WALKS AND DRIVEWAYS**			
	Bituminous - 1 1/2" with 4" base	SqFt	.40	.58
	2" with 4" base	SqFt	.46	.60
	Concrete - 4" - Broom Finish	SqFt	1.42	1.15
	5" - Broom Finish	SqFt	1.45	1.35
	6" - Broom Finish	SqFt	1.50	1.40
	Add Mesh 6" x 6", 10 - 10	SqFt	.10	.15
	Add for Exposed Aggregate	SqFt	.68	.15
	Add for Coloring (Acrylic)	SqFt	.35	.60
	Add for Base - 4" Sand or Gravel	SqFt	.25	.20
	Precast Block - Colored 1" with 2" Sand Cushion	SqFt	1.30	.80
	Colored 2" with 2" Sand Cushion	SqFt	1.50	1.00
	Crushed Rock - 4" Compacted	SqFt	.30	.28
	Flagstone - 1 1/4" with Sand Bed	SqFt	3.60	7.00
	1 1/4" with Mortar Setting Bed	SqFt	5.00	8.00
	Brick - 4" x 8" x 2 1/4" with 4" Sand Bed	SqFt	3.50	2.80
	4" x 8" x 2 1/4" with Mortar Setting Bed	SqFt	4.10	3.60
	Add for Herringbone or Weave Pattern	SqFt	1.25	1.35
	Wood - 2" T&G on 6" x 6" Timbers	SqFt	1.40	2.50
	2" Boards on 4" x 4" Timbers	SqFt	1.25	2.10
	Asphalt Block - 6" x 12" x 3"	SqFt	2.20	4.00
	Slate - 1 1/4"	SqFt	5.40	7.50
.4	**STAIRS - EXTERIOR**	SqFt	6.30	1.30

DIVISION #2 - SITEWORK & DEMOLITION

0207.0 FENCING (L&M) (Ironworkers or Carpenters)

	UNIT	COST
.1 CHAIN LINK (Galvanized 9 ga) Including Intermediate Posts and Concrete Embedded		
5' High - 2" Mesh and 2" O.D. Pipe	LnFt	12.50
6' High	LnFt	13.50
7' High	LnFt	14.00
8' High	LnFt	16.50
10' High	LnFt	31.00
Add for 1 - or 3 - Strand Barbed Wire	LnFt	2.70
Add for Vinyl Coated Wire	LnFt	15%
Add for Aluminum Wire	LnFt	25%
Add for Wood Slats	LnFt	50%
Add for Aluminum Slats	LnFt	75%
Add for Posts - Corner, End and Gate:		
5' High - 3" O.D.	Each	40.00
6' High - 3" O.D.	Each	45.00
7' High - 3" O.D.	Each	57.00
8' High - 3" O.D.	Each	66.00
10' High - 3" O.D.	Each	93.00
Add for Gate Frames:		
5' High	LnFt	39.00
6' High	LnFt	41.00
7' High	LnFt	52.00
8' High	LnFt	66.00
10' High	LnFt	80.00
.2 WELDED WIRE (Galvanized 11 ga) Including Intermediate Posts and Concrete Embedded		
4' High - 2" x 4"	LnFt	10.20
5' High	LnFt	11.50
6' High	LnFt	12.50
Add for Posts:		
4' Long	Each	22.50
5' Long	Each	24.00
6' Long	Each	26.00
Add for Gate Frames:		
4' Long	LnFt	21.50
5' Long	LnFt	26.00
6' Long	LnFt	28.00
Add for Slats	LnFt	5.00
.3 MESH WIRE (Galvanized 14 ga)		
3' High	LnFt	5.00
4' High	LnFt	6.00
5' High	LnFt	6.80
.4 WOOD FENCING		
Rail - 4' Cedar - 3 Boards	LnFt	14.50
Redwood - 3 Boards	LnFt	15.00
Pine, Painted - 3 Boards	LnFt	14.00
Vertical - 6' Pine, Painted	LnFt	17.50
Redwood	LnFt	21.50
Cedar	LnFt	22.50
Weave - 6' Redwood	LnFt	23.00
8' Redwood	LnFt	26.00
Picket - 4' Cedar	LnFt	11.50
6' Cedar	LnFt	14.50
Split Rail - 4' - 3 Rail	LnFt	7.80
Add for Post Set in Concrete	Each	5.00
.5 GUARD RAILS - Corrugated - Galvanized with Wood Posts	LnFt	28.00
Galvanized with Steel Posts	LnFt	30.00
Wire Cable - with Wood Posts, 3 - Strand	LnFt	14.50
with Steel Posts, 3 - Strand	LnFt	15.50
with no posts, 3-strand	LnFt	7.00

DIVISION #2 - SITEWORK & DEMOLITION

0208.0 RECREATIONAL FACILITIES

		UNIT	COST
.1	PLAYING FIELDS AND OUTDOOR COURTS		
.11	Football Fields - Field Size - 300' x 160		
	Minimum Use Size - 306' x 180'		
	Artificial Turf	Each	620,000.00
	Natural Turf	Each	280,000.00
.12	Tennis Courts -Court Size - 78' x 36'		
	Minimum Use Size - 108' x 48'		
	Asphalt with Color Surfacing - 1 Court	Each	23,000.00
	2 Courts	Each	18,500.00
	3 Courts	Each	18,000.00
	Incl. 3" Asphalt, 4" Gravel Base & Striping		
	Synthetic	Each	32,000.00
	Concrete	Each	27,000.00
	Add for Practice Boards	Each	2,000.00
	Add for Net Posts (4" Pipe with Ratchet)	Pair	550.00
	Add for Steel nets	Each	550.00
.13	Track and Field		
	Track - 440 yard - Track Size 21' x 1,320'		
	Synthetic Turf - 1" Polyurethane	Each	132,000.00
	1" Polyurethane	SqYd	37.00
	3" Bituminous Base	SqYd	7.00
	6" Gravel Base	SqYd	5.00
	Rubberized Asphalt	Each	62,000.00
	1" with 12% Resiliency	SqYd	10.50
	Add Bituminous Base - same as above	SqYd	6.50
	Cinder	Each	52,000.00
	High Jump - 50' Radius w/12' x 16' P.T.	Each	9,600.00
	Long Jump - 200' x 4' Runway w/20' x 8' P.T.	Each	4,200.00
	Pole Vault - 130' x 4' Runway w/16' x 16' P.T.	Each	2,700.00
.14	Basketball Courts -		
	Concrete - 4" Grade School - Court 42' x 74'	Each	16,000.00
	High School - Court 50' x 84'	Each	19,000.00
	College - Court 50' x 94'	Each	22,500.00
	Asphalt - 1 1/2" - Grade School	Each	7,800.00
	High School	Each	8,800.00
	College	Each	10,600.00
	Add for Back Stops	Each	1,300.00
.15	Volley Ball Courts - Court Size 30' x 60'	Each	9,500.00
	Minimum Use Size - 36' x 66'		
	Concrete - 4"	Each	8,500.00
	Asphalt - 1 1/2"	Each	6,600.00
	Add for Posts and Inserts	Pair	530.00
.16	Shuffleboard Courts - 6' x 52'		
	Concrete - 4"	Each	1,250.00
	Asphalt - 1 1/2"	Each	800.00
.17	Miniature Golf (9 - Hole) Bases - Concrete	Each	9,500.00
	Equipment and Carpet	Each	7,700.00
.18	Horse Shoe Courts - 10' x 40'	Each	540.00
	See Division 11 for Playground Equipment		
.19	Baseball and Softball Fields		
	45' Radius with Backstop	Each	12,000.00
	65' Radius with Backstop	Each	14,600.00

DIVISION #2 - SITEWORK & DEMOLITION

0208.0 RECREATIONAL FACILITIES, Cont'd...

	UNIT	COST
.2 RECREATIONAL EQUIPMENT		
Climbers - Arch	Each	450.00
Dome	Each	530.00
Circular	Each	1,020.30
Catwalks	Each	650.00
Slides - Straight - 6'	Each	870.00
8'	Each	1,100.00
10'	Each	1,650.00
Spiral - 7'	Each	4,000.00
10'	Each	6,100.00
Spring Equipment - Single	Each	380.00
2 Unit	Each	575.00
4 Unit	Each	1,060.00
Swings - 2 Leg - 2 Seat - 8' High	Each	640.00
2 Leg - 4 Seat	Each	925.00
3 Leg - 3 Seat - 8' High	Each	710.00
6 Seat	Each	1,060.00
Add for 12' High	-	20%
Add for Nursery Seats	Each	255.00
Add for Saddle Seats	Each	590.00
Add for Tire Swings	Each	150.00
Teeter Totters	Each	950.00
Whirls - Small - 6'	Each	780.00
Large - 10'	Each	1,350.00
.3 SITE FURNISHINGS		
Benches - Steel Leg		
Stationary - w/Black Wood Slats - 8'	Each	390.00
w/Black Aluminum Slats - 8'	Each	525.00
Portable - 8'	Each	250.00
Player - No Back, Wood Seat - 8'	Each	205.00
No Back, Wood Seat - 14'	Each	330.00
Bike Racks		
Permanent - 10'	Each	530.00
20'	Each	820.00
Portable - 10'	Each	300.00
Bleachers, Elevated		
5 - Row x 15' Steel - Seats 50	Each	3,300.00
10 - Row x 15' Steel - Seats 100	Each	5,700.00
15 - Row x 15' Steel - Seats 150	Each	8,000.00
Litter Receptacles		
Pedestal - 24"	Each	300.00
36"	Each	370.00
Stoves	Each	360.00
Tables - Steel Frame - Wood Slat - 8'	Each	380.00
Aluminum Slat - 8'	Each	420.00
All Wood	Each	250.00
All Aluminum	Each	430.00
.4 SHELTERS	SqFt	14.00

DIVISION #2 - SITEWORK & DEMOLITION

		UNIT	COST	UNIT	COST
0209.0	**LANDSCAPING (L&M) (Laborers)**				
.1	SEEDING (.80 lb) - Machine	SqYd	.40	Acre	1,750.00
	Hand	SqYd	.90	Acre	4,300.00
	Add Fine Grading - Hand	SqYd	.80	Acre	3,700.00
	Add for 30 - Day Maintenance	-	-	Acre	1,250.00
	Deduct for Rye and Wild Grasses	-	-	Acre	560.00
.2	SODDING - Flat ($.80 per SqYd)	SqYd	2.35	-	-
	Slope (Pegged)	SqYd	2.90	-	-
	Add Fine Grading - Hand	SqYd	.90	-	-
	Add for 30 - Day Maintenance	SqYd	.40	-	-
	Remove and Haul Away Old Sod	SqYd	1.05		
.3	DISCING	SqYd	.85	Acre	4,100.00
.4	FERTILIZING	SqYd	.25	Acre	1050.00
				Ton	850.00
.5	TOP SOIL - 4" Hand - Only	SqYd	3.00	Acre	16,000.00
	Machine and Hand	SqYd	2.10	Acre	10,400.00
		CuYd	27.00	-	-
.6	TREES AND SHRUBS				
	Trees - 2"			Each	410.00
	3"			Each	570.00
	4"			Each	770.00
	5"			Each	970.00
	6"			Each	1,400.00
	Shrubs - Small - 1' to 3'			Each	20.00
	Large - 3' to 5'			Each	40.00
.7	BEDS				
	Wood Chips, Cedar, 2" - $12 CuYd			SqYd	2.70
	Mulch, Redwood Bark - $30 CuYd			SqYd	5.65
	Stone Aggregate - $15 CuYd			SqYd	3.10
	Planting Soil, 24" - $15 CuYd			SqYd	3.50
.8	WOOD CURBS AND WALLS				
	Curbs - 4" x 4" - Fir or Pine Treated			LnFt	2.30
	Cedar			LnFt	3.00
	6" x 6" - Fir or Pine Treated			LnFt	6.00
	Cedar			LnFt	7.50
	Walls - 4" x 4" - Fir or Pine, Incl. Dead Men			SqFt	6.25
	6" x 6" - Fir or Pine, Incl. Dead Men			SqFt	9.30
	Cedar, Including Dead Men			SqFt	14.20
.9	SPRINKLER SYSTEM, WATERED AREA			CSF	32.00
0210.0	**SOIL STABILIZATION**				
.1	CHEMICAL INTRUSION			CuFt	15.00
	Pressure Grout			Gal	4.75
.2	CONCRETE INTRUSION			CuFt	28.00
.3	VIBRO FLOTATION			-	-
0211.0	**SOIL TREATMENT**				
.1	PESTICIDE			SqFt	.25
.2	VEGETATION PREVENTATIVE			SqFt	.10
0212.0	**RAILROAD WORK**				
.1	NEW WORK	LnFt			125.00
.2	REPAIR WORK			LnFt	18.50
0213.0	**MARINE WORK**				
0214.0	**TUNNEL WORK**				

DIVISION #2 - SITEWORK & DEMOLITION - QUICK ESTIMATING

0201.0 DEMOLITION

.32 Selective Building Removals -
No Cutting or Disposal Included (see 0201.34 & 0201.35)

	UNIT	COST
<u>Concrete</u> - Hand Work (see 0201.32 for Machine Work)		
8" Walls - Reinforced	SqFt	7.50
Non - Reinforced	SqFt	6.00
12" Walls - Reinforced	SqFt	18.00
12" Footings x 24" wide	LnFt	15.50
x 36" wide	LnFt	21.50
16" Footings x 24" wide	LnFt	27.00
6" Structural Slab - Reinforced	SqFt	6.20
8" Structural Slab - Reinforced	SqFt	7.50
4" Slab on Ground - Reinforced	SqFt	2.90
Non - Reinforced	SqFt	2.10
6" Slab on Ground - Reinforced	SqFt	3.40
Non - Reinforced	SqFt	2.50
Stairs - Reinforced	SqFt	10.80
<u>Masonry</u> - Hand Work		
4" Brick or Stone Walls	SqFt	2.05
4" Brick and 8" Backup Block or Tile	SqFt	3.70
4" Block or Tile Partitions	SqFt	1.90
6" Block or Tile Partitions	SqFt	2.00
8" Block or Tile Partitions	SqFt	2.35
12" Block or Tile Partitions	SqFt	3.00
<u>Miscellaneous</u> - Hand Work (Including Loading)		
Acoustical Ceilings - Attached (Including Iron)	SqFt	.77
Suspended (Including Grid)	SqFt	.55
Asbestos - Pipe	LnFt	70.00
Ceilings and Walls	SqFt	22.00
Columns and Beams	SqFt	55.00
Tile Flooring	SqFt	2.60
Cabinets and Tops	LnFt	15.00
Carpet	SqFt	.44
Ceramic and Quarry Tile	SqFt	1.70
Doors and Frames - Metal	Each	76.00
Wood	Each	68.00
Drywall Ceilings - Attached	SqFt	1.10
Drywall on Wood or Metal Studs - 2 Sides	SqFt	1.15
Paint Removal - Doors and Windows	SqFt	1.15
Walls	SqFt	.90
Plaster Ceilings - Attached (Including Iron)	SqFt	1.60
on Wood or Metal Studs	SqFt	1.65
Roofing - Builtup	SqFt	1.50
Shingles - Asphalt and Wood	SqFt	.48
Terrazzo Flooring	SqFt	2.50
Vinyl Flooring	SqFt	.51
Wall Coverings	SqFt	.95
Windows	Each	50.00
Wood Flooring	SqFt	.52

.33 Site Removals (Including Loading)

	UNIT	COST
4" Concrete Walks - Labor Only (Non-Reinforced)	SqFt	2.10
Machine Only (Non-Reinforced)	SqFt	.70
6" Concrete Drives - Labor Only (Non-Reinforced)	SqFt	2.70
Machine Only (Reinforced)	SqFt	1.00
6" x 18" Concrete Curb - Machine	LnFt	1.80
Curb and Gutter - Machine	LnFt	2.40
2" Asphalt - Machine	SqFt	.45
Fencing - 8' Hand	LnFt	2.60

DIVISION #2 - SITEWORK & DEMOLITION - QUICK ESTIMATING

		UNIT	COST
0202.0	**EARTHWORK**		
.1	GRADING - Hand - 4" - Site	SqFt	.23
	4" - Building	SqFt	.40
.2	EXCAVATION - Hand - Open - Soft (Sand)	CuYd	33.00
	Medium (Clay)	CuYd	45.00
	Hard (Shale)	CuYd	70.00
	Add for Trench or Pocket		15%
.3	BACK FILL - Hand - Not Compacted (Site Borrow)	CuYd	16.50
.4	BACK FILL - Hand - Compacted (Site Borrow)		
	12" Lifts - Building - No Machine	CuYd	27.00
	With Machine	CuYd	22.00
	18" Lifts - Building - No Machine	CuYd	24.00
	With Machine	CuYd	18.00
0205.0	**DRAINAGE**		
.4	BUILDING FOUNDATION DRAINAGE		
	4" Clay Pipe	LnFt	3.70
	4" Plastic Pipe - Perforated	LnFt	2.80
	6" Clay Pipe	LnFt	4.20
	6" Plastic Pipe - Perforated	LnFt	4.40
	Add for Porous Surround - 2' x 2'	LnFt	5.50
0206.0	**PAVEMENT, CURBS AND WALKS**		
.2	CURBS AND GUTTERS		
.21	Concrete - Cast in Place (Machine Placed)		
	Curb - 6" x 12"	LnFt	13.00
	6" x 18"	LnFt	15.50
	6" x 24"	LnFt	21.00
	6" x 30"	LnFt	22.50
	Curb and Gutter - 6" x 12"	LnFt	16.00
	6" x 18"	LnFt	20.00
	6" x 24"	LnFt	23.50
	Add for Hand Placed	LnFt	8.00
	Add for 2 #5 Reinf. Rods	LnFt	2.20
	Add for Curves and Radius Work	LnFt	40%
.22	Concrete Precast - 6" x 10" x 8"	LnFt	9.60
	6" x 9" x 8"	LnFt	9.00
.23	Bituminous - 6" x 8"	LnFt	2.80
.24	Granite - 6" x 16"	LnFt	33.00
.25	Timbers - Treated - 6" x 6"	LnFt	6.70
	6" x 8"	LnFt	9.25
.26	Plastic - 6" x 6"	LnFt	8.50
.3	WALKS		
	Bituminous - 1 1/2" with 4" Sand Base	SqFt	1.35
	2" with 4" Sand Base	SqFt	1.40
	Concrete - 4" - Broom Finish	SqFt	4.10
	5" - Broom Finish	SqFt	4.20
	6" - Broom Finish	SqFt	4.40
	Add for 6" x 6", 10 - 10 Mesh	SqFt	.30
	Add for 4" Sand Base	SqFt	.55
	Add for Exposed Aggregate	SqFt	1.35
	Crushed Rock - 4"	SqFt	.65
	Brick - 4" - with 2" Sand Cushion	SqFt	9.30
	4" - with 2" Mortar Setting Bed	SqFt	10.50
	Flagstone - 1¼" - with 4" Sand Cushion	SqFt	14.30
	1¼" - with 2" Mortar Setting Bed	SqFt	18.00
	Precast Block - 1" Colored with 4" Sand Cushion	SqFt	3.30
	2" Colored with 4" Sand Cushion	SqFt	4.00
	Wood - 2" Boards on 6" x 6" Timbers	SqFt	5.25
	2" Boards on 4" x 4" Timbers	SqFt	4.60
	Slate - 1¼" - with 2" Mortar Setting Bed	SqFt	19.00

DIVISION #3 - CONCRETE

Wage Rates (Including Fringes) & Location Modifiers
July 2002-2003

	Metropolitan Area		Placing Labor Rate		Cement Finisher Rate		Carpenter Rate	Wage Rate Location Modifier
1.	Akron	*	27.82	*	34.29	*	33.59	114
2.	Albany-Schenectady-Troy		26.27		33.00		29.62	102
3.	Atlanta		14.03		21.08		22.05	80
4.	Austin	**	15.95	**	20.86	**	22.03	74
5.	Baltimore		18.36		26.16		26.15	92
6.	Birmingham	**	17.84	**	23.08	**	22.02	75
7.	Boston		31.50		43.04		39.47	136
8.	Buffalo-Niagara Falls		31.98		38.20		38.46	128
9.	Charlotte	**	13.96	**	19.32	**	20.07	67
10.	Chicago-Gary	*	32.95	*	39.43	*	39.15	126
11.	Cincinnati		24.69		26.51		27.92	95
12.	Cleveland		30.26		34.68		33.90	117
13.	Columbus	*	23.54	*	28.37	*	28.77	99
14.	Dallas-Fort Worth	**	14.39	**	21.47	**	20.00	68
15.	Dayton	*	24.22	*	30.75	*	31.34	105
16.	Denver-Boulder		16.39		24.36		24.12	81
17.	Detroit		30.71		36.31		38.25	127
18.	Flint	*	25.33	*	30.64	*	30.41	100
19.	Grand Rapids		23.03		29.84		30.40	100
20.	Greensboro-West Salem	**	13.98	**	19.26	**	20.18	67
21.	Hartford-New Britain	*	28.60	*	36.30	*	34.29	116
22.	Houston	**	16.93	**	22.07	**	23.14	73
23.	Indianapolis		24.08		32.74		31.29	104
24.	Jacksonville	**	16.05	**	23.64	**	24.63	82
25.	Kansas City		24.25		30.09		29.79	98
26.	Los Angeles-Long Beach		30.23		35.76		33.06	122
27.	Louisville	**	19.38	**	25.18	**	24.40	82
28.	Memphis	**	17.71	**	21.41	**	20.77	87
29.	Miami	**	16.90	**	22.29	**	22.27	75
30.	Milwaukee	*	29.62	*	32.98	*	34.50	112
31.	Minneapolis-St. Paul		30.52		35.15		34.63	113
32.	Nashville	**	15.56	**	20.23	**	23.68	79
33.	New Orleans	**	14.60	**	18.14	**	21.64	73
34.	New York		42.64		52.05		55.76	193
35.	Norfolk-Portsmouth	**	15.67	**	20.25	**	23.28	79
36.	Oklahoma City	**	16.61	**	24.37	**	22.27	75
37.	Omaha-Council Bluffs		20.25		23.19		23.18	75
38.	Orlando	**	19.10	**	25.64	**	25.97	89
39.	Philadelphia	*	32.57	*	39.08	*	40.85	135
40.	Phoenix	**	18.09	**	22.76	**	24.84	95
41.	Pittsburgh	*	24.73	*	30.87	*	32.53	109
42.	Portland	*	28.70	*	34.32	*	33.58	113
43.	Providence-Pawtucket	*	27.70	*	33.06	*	33.70	114
44.	Richmond	**	15.56	**	19.77	**	23.23	78
45.	Rochester		27.07		35.06		35.02	116
46.	Sacramento	*	31.01	*	38.43	*	40.67	132
47.	St. Louis		30.32		34.46		35.63	117
48.	Salt Lake City-Ogden	**	17.34	**	24.30	**	23.05	70
49.	San Antonio	**	12.70	**	18.58	**	21.61	75
50.	San Diego	*	31.09	*	28.77	*	33.43	124
51.	San Francisco-Oakland-San Jose		31.04		33.75		40.67	130
52.	Seattle-Everett		30.74		37.48		34.95	116
53.	Springfield-Holyoke-Chicopee	*	32.16	*	31.27	*	32.71	110
54.	Syracuse	*	24.87	*	28.53	*	28.62	114
55.	Tampa-St. Petersburg	**	17.78	**	26.28	**	26.68	91
56.	Toledo	*	26.10	*	34.13	*	34.49	117
57.	Tulsa	**	16.57	**	24.37	**	22.70	75
58.	Tucson	**	20.82	**	28.44	**	28.14	95
59.	Washington D.C.	*	17.60	*	25.79	*	24.77	87
60.	Youngstown-Warren	*	27.74	*	31.77	*	30.45	107
	AVERAGE		**23.30**		**29.34**		**29.48**	

Note: See Division 5 for Reinforcing Steel Location Modifiers and Wage Rates
* Contract Not Settled - Wage Interpolated
Impact Ratio: Placing   Labor 20%   Material 80%
               Finishing   Labor 92%   Material 8%
               Form Work   Labor 75%   Material 25%
               Reinforcing Steel Labor 35% Material 65%
** Non Signatory or Open Shop Rate

DIVISION #3 - CONCRETE

		PAGE
0301.0	**CONCRETE - Cast in Place (CSI 03300)**	3-4
.1	CONCRETE PLACING	3-4
.11	Footings	3-4
.12	Walls & Grade Beams	3-4
.13	Columns and Pedestals	3-4
.14	Beams	3-4
.15	Slabs (Structural)	3-4
.16	Stairs and Landings	3-4
.17	Curbs, Platforms and Miscellaneous Small Pours	3-4
.18	Slabs on Ground	3-4
.19	Slabs Over Decks or Lath	3-4
.20	Toppings	3-4
.2	CONCRETE FINISHING (INCLUDING SCREEDS)	3-5
.21	Rough Screeding	3-5
.22	Trowel Finishing	3-5
.23	Brush or Broom Finishing	3-5
.24	Float Finishing	3-5
.25	Special Finishing (Rub, Patch, Retard, Sandblasts, etc)	3-5
.3	SPECIALTIES	3-6
.31	Abrasives	3-6
.32	Admixtures	3-6
.33	Colors	3-6
.34	Curing	3-6
.35	Expansion and Control Joints	3-6
.36	Grouts	3-7
.37	Hardeners and Sealers	3-7
.38	Joint Sealers	3-7
.39	Moisture Proofing	3-7
.4	EQUIPMENT	3-7
.5	TESTS	3-7
0302.0	**FORMWORK (CSI 03100)**	3-7
.1	REMOVABLE FORMS (Expendable and Reusable)*	3-7
.11	Footings	3-7
.12	Walls and Grade Beams	3-7
.13	Columns	3-8
.14	Beams	3-8
.15	Slabs (Flat and Pan)	3-9
.16	Stairs and Landings	3-10
.17	Curbs and Platforms	3-10
.18	Edge Forms and Bulkheads for Slabs	3-10
.2	SHORING, SUPPORTS AND ACCESSORIES	3-10
.21	Horizontal Shoring (Beam and Joist)	3-10
.22	Vertical Shoring (Posts and Tubular Frames)	3-10
.23	Column Clamps and Steel Strapping	3-10
.24	Wall, Column and Beam Forms	3-10
.25	Accessories (Hangers, Tyloops, Lags, etc.)	3-11
.3	SPECIALTIES	3-11
.31	Nails, Wire and Ties	3-11
.32	Anchors and Inserts	3-11
.33	Chamfers, Drips, etc	3-11
.34	Stair Nosings and Treads	3-12
.35	Water Stops	3-12
.36	Divider Strips	3-12
.37	Tongue and Groove Joint Forms	3-12
.38	Bearing Pads and Shims	3-12
.4	EQUIPMENT	3-12

* See Section 0505 for Permanent Forms of Corrugated Deck and Steel Lath.

DIVISION #3 - CONCRETE

<u>PAGE</u>

0303.0 REINFORCING STEEL (CSI 03200) — **3-13**
- .1 BARS — 3-13
- .2 ACCESSORIES — 3-13
- .3 POST TENSION STEEL (Cable and Wire) — 3-13
- .4 WIRE FABRIC — 3-14

0304.0 SPECIALTY PLACED CONCRETE — **3-14**
- .1 LIFT SLAB — 3-14
- .2 POST TENSIONED OR STRESSED IN PLACE — 3-14
- .3 TILT UP PANELS
- .4 PNEUMATICALLY PLACED — 3-14

0305.0 PRECAST PRESTRESSED CONCRETE — **3-14**
- .1 BEAMS AND COLUMNS — 3-14
- .2 DECKS (Flat) — 3-14
- .3 SINGLE AND DOUBLE T'S — 3-14
- .4 WALLS — 3-14

0306.0 PRECAST CONCRETE (CSI 03400) — **3-15**
- .1 BEAMS AND COLUMNS — 3-15
- .2 DECKS — 3-15
- .3 PANELS (Wall and Facing Units) — 3-15
- .4 PLANK — 3-15
- .5 SPECIALTIES (Curbs, Copings, Sills, Stools, Medians, etc.) — 3-15

0307.0 PRECAST CEMENTITIOUS PLANK (CSI 03500) — **3-15**
- .1 EXPANDED MINERALS — 3-15
- .2 GYPSUM (CSI 03510) — 3-15
- .3 WOOD FIBRE (CSI 03530) — 3-15

0308.0 POURED IN PLACE CEMENTITIOUS CONCRETE — **3-15**
- .1 EXPANDED MINERALS — 3-15
- .2 EXPANDED MINERALS AND ASPHALT — 3-15
- .3 GYPCRETE — 3-15

3A   QUICK ESTIMATING — **3-16**

3-17

3-18

DIVISION #3 - CONCRETE

0301.0 GENERAL CONCRETE WORK (S) (Cement Finishers & Laborers)

		UNIT	LABOR	MATERIAL
.1	CONCRETE PLACING & VIBRATING (3500# Concrete w/5.5 sack cement)			
.11	Footings - 1 1/2" Aggregate			
	Wall & Pad Types - Truck Chuted	CuYd	12.80	75.00
	Buggies	CuYd	16.50	75.00
	Crane	CuYd	17.50	75.00
.12	Wall & Grade Beams - 3/4" Aggregate			
	At Grade (to 8' deep) -Truck Chuted	CuYd	13.40	78.00
	Buggies	CuYd	18.00	78.00
	Crane	CuYd	18.00	78.00
	At Grade (over 8' deep, poured with trunks) - Truck Chuted	CuYd	14.30	78.00
	Buggies	CuYd	19.10	78.00
	Crane & Hoppers	CuYd	18.00	78.00
	Above Grade (deeper than 8') - Conveyors	CuYd	13.30	78.00
	Ramp & Buggies	CuYd	19.10	78.00
	Crane & Hoppers	CuYd	18.00	78.00
	Climbing Crane	CuYd	16.90	78.00
	Pumping	CuYd	14.30	78.00
.13	Column & Pedestals - 3/4" Aggregate - Buggies	CuYd	21.20	78.00
	Crane	CuYd	18.00	78.00
	Tower & Buggies	CuYd	20.10	78.00
	Fork Lift	CuYd	37.00	78.00
	Climbing Crane	CuYd	19.10	78.00
	Pumping	CuYd	15.40	78.00
.14	Beams - 3/4" Aggregate - Buggies	CuYd	20.10	78.00
	Crane	CuYd	18.00	78.00
	Tower & Buggies	CuYd	20.00	78.00
	Fork Lift	CuYd	37.00	78.00
	Climbing Crane	CuYd	18.00	78.00
	Pumping	CuYd	16.00	78.00
.15	Slabs - 6" Structure - 3/4" Aggregate - Buggies	CuYd	17.10	78.00
	Crane	CuYd	18.00	78.00
	Climbing Crane	CuYd	18.00	78.00
	Pumping	CuYd	15.50	78.00
.16	Stairs & Landings -			
	Structural - 3/4" Aggregate - Buggies	CuYd	20.20	78.00
	Truck Chuted	CuYd	16.00	78.00
	Climbing Crane	CuYd	18.00	78.00
	Pumping	CuYd	16.00	78.00
	Pan Filled - 1/2" Ready Mixed	CuYd	35.00	83.00
	Hand Mixed (Dry)	CuYd	54.80	70.00
.17	Curbs, Platforms & Misc. Small Pours - 1/2" Aggregate Buggies	CuYd	25.30	83.00
	Fork Lift	CuYd	35.00	83.00
.18	Slabs on Ground - 3/4" Aggregate - Truck Chuted	CuYd	13.80	78.00
	Buggies	CuYd	18.00	78.00
	Crane	CuYd	17.00	78.00
	Conveyors	CuYd	14.30	78.00
.19	Slabs-Over Metal Decks or Lath - 1/2" Aggregate Buggies	CuYd	21.20	83.00
	Tower & Buggies	CuYd	22.40	83.00
	Crane	CuYd	19.20	83.00
.20	Toppings - 1/2" Aggregate - Buggies	CuYd	20.20	83.00
	Towers & Buggies	CuYd	22.30	83.00

Conveying Equipment and Operator not included in above costs. See Page 3-5. Also other variations on Page 3-5

DIVISION #3 - CONCRETE

0301.0	GENERAL CONCRETE WORK, Cont'd...	UNIT	LABOR	MATERIAL
	Additions to Concrete Placing Work:			
	Add for Ea 500# Concrete above 3500# Concrete	CuYd	-	2.50
	Add or deduct for Sack Cement	CuYd	-	5.20
	Add for High Carbon Concrete	CuYd	-	18.50
	Add for High Early Cement Concrete	CuYd	-	8.00
	Deduct for 1 1/2" Aggregate	CuYd	-	1.75
	Add for 3/8" or 1/2" Aggregate	CuYd	-	3.25
	Add for Lightweight Aggregate (4000#)	CuYd	2.65	16.00
	Add for Heavyweight Aggregate (Granite) (4000#)	CuYd	5.35	11.00
	Deduct for Fly Ash Concrete (3000#)	CuYd	-	1.00
	Add for Exposed Aggregate Mix	CuYd	-	8.50
	Add for Fibre Reinforcement Mix	CuYd	3.20	8.50
	Add for Less than 6 CuYd Delivery	Load	-	45.00
	Add for Hauls beyond 15 Miles	CuYd Mi	-	1.00
	Add for Time after 7 Minutes per Yard	Minute	-	.80
	Add for Conveying Equipment and Operators:			
	Towers (Approx. 30 CuYd/ Hour)	CuYd	-	6.00
	Mobile Cranes (Approx. 25 CuYd/ Hour)	CuYd	-	7.00
	Tower Cranes (Approx. 20 CuYd/ Hour)	CuYd	-	7.50
	Fork Lifts (Approx. 5 CuYd/ Hour)	CuYd	-	7.00
	Conveyors (Approx. 50 CuYd/ Hour)	CuYd	-	7.50
	Pumping (Approx. 35 CuYd/ Hour)	CuYd	-	7.50
	Add for Floors Above Grade - per Floor Placing	CuYd	1.00	-
	Add for Winter Work:			
	Productivity Loss	CuYd	4.30	-
	Heating Water and Aggregate	CuYd	-	4.50
	Calcium Chloride 1%	CuYd	-	2.20
	Insulation Blanket - Slabs (5 uses)	SqFt	.08	.13
	Walls & Beams (3 uses)	SqFt	.10	.10
	Heaters: Without Operators (Including Fuel - Floor Area)	CuYd	.38	1.50
		or SqFt	.04	.12
	Enclosures (3 uses)- Wall Area	CuYd	.86	.35
		or SqFt	.08	.15
.2	FINISHING (Including Screeds and Bulkheads to 6")			
.21	Rough Screeding - Slabs	SqFt	.28	.07
	Stairs	SqFt	.49	.07
.22	Trowel Finishing - Slabs on Ground	SqFt	.42	.07
	Solid and Pan Slabs	SqFt	.45	.07
	Slab over Corrugated	SqFt	.51	.07
	Topping Slabs	SqFt	.45	.07
	Stairs	SqFt	.78	.07
	Curbs and Bases	SqFt	.78	.07
.23	Broom Finishing - Slabs on Ground	SqFt	.38	.07
	Solid and Pan Slabs	SqFt	.43	.07
	Slab over Corrugated	SqFt	.51	.07
	Topping Slabs	SqFt	.43	.07
	Stairs	SqFt	.78	.07
.24	Float Finish - Slabs on Ground	SqFt	.36	.07
	Solid and Pan Slabs	SqFt	.40	.07
	Slab over Corrugated	SqFt	.43	.07
	Stairs	SqFt	.70	.07
.25	Special Finishing			
.251	Patch Walls, 2 Sides Tie Holes & Honeycomb	SqFt	.15	.07
.252	Rub Walls, 1 Side Carborundum for Fins	SqFt	.09	.03
	With Burlap and Grout	SqFt	.45	.07
.253	Level & Top Floors - Trowel Finish 1"	SqFt	.78	.35
	With Epoxy 1/4"	SqFt	1.55	5.00
.254	Exposed Aggregate Slabs - Washed	SqFt	.59	.05
	Retardant	SqFt	.20	.10
	Add for - Seeding	SqFt	.30	.10
	Add to All Finishing Items Above			
	Winter Production Loss and Cost	SqFt	.15	.05
	Sloped Work	SqFt	.12	.04
	Heavyweight Aggregates	SqFt	.15	.05

DIVISION #3 - CONCRETE

		UNIT	LABOR	MATERIAL
0301.0	**GENERAL CONCRETE WORK, Cont'd...**			
.255	Bushhammer - Green Concrete	SqFt	1.03	.08
	Cured Concrete	SqFt	1.55	.10
	Sand Blast - Light Penetration	SqFt	.75	.10
	Heavy Penetration	SqFt	1.30	.20
.3	SPECIALTIES			
.31	Abrasives - Non-Slip			
	Alo-Grit (.86 lb & 1/4 lb/SqFt)	SqFt	.19	.23
	Carborundum Grits (1.30 lb & 1/4 lb/SqFt)	SqFt	.19	.38
	Strips 3/8" x 1/4"	LnFt	.68	1.35
	Epoxy Coating (50.00 gal) & 50 SqFt/Gal	SqFt	.27	1.45
.32	Admixtures (per CuYd Concrete)			
	Accelerator (.06 oz)	CuYd	-	1.40
	Air Entraining (.04 oz)	CuYd	-	.80
	Densifiers (.07 oz)	CuYd	-	1.75
	Retarders (.05 oz)	CuYd	-	.80
	Water Reducing (.06 oz)	CuYd	-	2.00
.33	Colors			
	Dust on Type:			
	Black, Brown & Red (.50 lb) 50#/100 SqFt	SqFt	.19	.34
	Green and Blue (.60 lb) 50#/100 SqFt	SqFt	.19	.42
	Integral (Top 1"):			
	Black, Brown & Red (1.50 lb) 27#/100 SqFt	SqFt	.34	.60
	Green and Blue (4.25 lb) 27#/100 SqFt	SqFt	.34	1.30
	Full Thickness			
	Black, Brown & Red	CuYd	21.00	115.00
	Green and Blue	CuYd	21.00	150.00
.34	Curing			
	Curing Compounds			
	Resin (10.00 gal) 200 SqFt/gal	SqFt	.08	.06
	Hydrocide Res.Base (15.00 gal) 200 SqFt/gal	SqFt	.07	.08
	Rubber Base (8.00 gal) 200 SqFt/gal	SqFt	.07	.05
	Wax Base (3.00 gal) 200 SqFt/gal	SqFt	.07	.03
	Asphalt Base (4.00 gal) 200 SqFt/gal	SqFt	.07	.03
	Paper	SqFt	.08	.07
	Polyethylene - 4 mil	SqFt	.08	.05
	Water	SqFt	.08	.01
	Burlap	SqFt	.09	.07
	Curing and Sealing	SqFt	.10	.07
.35	Expansion and Control Joints			
	Asphalt - Fibre - 1/2" x 4"	LnFt	.32	.22
	1/2" x 6"	LnFt	.33	.32
	1/2" x 8"	LnFt	.40	.42
	Polyethylene Foam - 1/2" x 4"	LnFt	.32	.37
	1/2" x 6"	LnFt	.34	.47
	1/2" x 8"	LnFt	.40	.58
	Sponge Rubber - 1/2" x 4"	LnFt	.32	2.00
	1/2" x 6"	LnFt	.34	3.00
	1/2" x 8"	LnFt	.40	4.00
	Paper - Fibre (No Oil) - 1/2" x 4"	LnFt	.32	.26
	1/2" x 6"	LnFt	.34	.36
	1/2" x 8"	LnFt	.40	.45
	Add for Cap	LnFt	.18	.28
	Add for 3/4" Thickness	LnFt	-	30%
	Add for 1" Thickness	LnFt	-	80%
	See 0413.5 for Other Expansion Joint Costs.			
	See 0410.0 Mortars for Cement Prices.			

DIVISION #3 - CONCRETE

0301.0 GENERAL CONCRETE WORK, Cont'd...

	UNIT	LABOR	MATERIAL
.36 Grouts - Non-Shrink			
Iron Oxide (.45/lb)			
Hand Mixed (1 Cem: 1 Sand: 1 Iron Oxide)	CuFt	5.90	26.00
Per 1"	SqFt	.75	2.10
Premixed (.45/lb)	CuFt	2.15	38.00
Aluminum Oxide (.45/lb)	CuFt	4.25	12.00
Non-Metallic, Premixed (.35/lb)	CuFt	3.15	28.00
.37 Hardeners and Sealers			
Acrylic Sealer (7.50 gal) 300 SqFt/gal	SqFt	.10	.05
Epoxy Sealer (25.00 gal) 300 SqFt/gal	SqFt	.10	.13
Urethane Sealer (9.50 gal) 400 SqFt/gal	SqFt	.10	.05
Liquid Hardeners (4.85 gal) 200 SqFt/gal	SqFt	.10	.05
.38 Joint Sealers			
Rubber Asphalt			
Hot 1/2" x 1/2" Joint (.70 lb)	LnFt	.52	.48
Cold 1/2" x 1/2" Joint (.54 lb)	LnFt	.37	.55
Epoxy (20.00 Qt)	LnFt	.52	1.40
.39 Moisture Proofing (Loose Laid for Slabs)			
Polyethylene 4 Mil	SqFt	.07	.03
6 Mil	SqFt	.07	.04
Asphalted Paper	SqFt	.07	.05
.4 EQUIPMENT (Conv., Finishing, Vibrating, etc.)			
See Divisions I-6A and I-7A			
.5 TESTS			
Cylinder - 6" x 12" (7-day and 28-day)	Ea	-	38.00
Add per Pickup	Ea	-	37.00

0302.0 CONCRETE FORM WORK (S) (Carpenters)
(Lumber @ 490.00 MBF & 3/4" BB Plywood @ 1.00 SqFt)
(Costs Include Erection, Stripping, Cleaning, Oiling)

	UNIT	LABOR	MATERIAL
.1 REMOVABLE FORMS (Expendable and Reusable)			
.11 Footings (2-1/2 BdFt/SqFt and 3 Uses)			
Wall Type - Constant Elevation	SqFt	2.00	.57
Pad Type	SqFt	2.10	.57
Add for Volume Elevations and Changes	SqFt	.15	-
Add for Deep Foundation (8' or more)	SqFt	.15	-
Add for Each Foot Form Deeper than 12"	SqFt	.09	.05
Add for Keyway	LnFt	.40	.12
.12 Walls & Grade Beams (2-1/2 BdFt/SqFt & 3 Uses)			
Straight Walls - 4' High	SqFt	2.05	.68
Add for Each Foot Higher than 4'	SqFt	.05	.04
Add for Walls Above Grade	SqFt	.06	-
Add for Deep Foundation (8' or more)	SqFt	.10	-
Add for Pilastered Wall, 24' O.C. Total Area	SqFt	.09	.03
Add for Pilastered Wall, Pilasters Only	SqFt	4.05	.58
Add for Openings - Opening Area	SqFt	2.65	.70
Add for Retaining and Battered Type Wall	SqFt	.80	.10
Add for Curved Wall	SqFt	1.60	.13
Add for Brick Ledge	LnFt	1.55	.14
Add for Parapet Wall - Hung	SqFt	2.40	.25
Add for Pit or Small Trench Walls	SqFt	1.45	.08
Add for Lined Forms - Hardwood Panels	SqFt	.35	.45
Gang Formed Walls - Make Up	SqFt	3.70	5.80
Move	SqFt	.85	.18
3/4" B-B Plywood - Avg - 5 Ply - 8 Uses @ 1.00	SqFt	-	-
7 Ply - 8 Uses @ 1.15	SqFt	-	-
M.D.O. Plywood - Avg. - 20 Uses @ 1.35	SqFt		

DIVISION #3 - CONCRETE

0302.0 CONCRETE FORM WORK, Cont'd...

	UNIT	LABOR	MATERIAL
.13 Columns & Pedestals			
Sq. & Rectangular (2 1/2 BdFt/ SqFt & 3 Uses)			
8" x 8"	SqFt	4.00	.82
12" x 12"	SqFt	3.95	.83
16" x 16"	SqFt	3.90	.84
20" x 20"	SqFt	3.75	.85
24" x 24"	SqFt	3.80	.86
Add per Foot - Work over 10' Floor Heights	SqFt	.32	-
Add per Floor above 20' Above Grade	SqFt	.12	-
Deduct for Ganged Formed (8 Uses)	SqFt	1.10	-
Round - Steel or Fiberglass (Rent and 3 Uses)			
12"	LnFt	6.15	4.00
16"	LnFt	6.00	4.40
20"	LnFt	7.10	5.20
24"	LnFt	10.30	6.00
30"	LnFt	13.40	7.25
36"	LnFt	16.70	9.10
Add per Foot - Work over 10" Floor Heights	LnFt	.32	.40
Round-Fibre (6" to 48" x 18' available)			
8" Cut Length Prices Used	LnFt	4.30	2.20
12"	LnFt	4.50	4.20
16"	LnFt	4.90	6.75
20"	LnFt	6.15	9.00
24"	LnFt	8.30	12.00
30"	LnFt	9.70	15.00
36"	LnFt	12.00	17.00
Add for Less than 100 Feet - One Size	LnFt	-	20%
Add for Seamless Fibre	LnFt	-	15%
Add for Conical Heads to Steel or Fibre	Each	85.00	55.00
Add for Beam Fittings or Other Openings	Each	95.00	60.00
.14 Beams (12" Floor Heights and 3 Uses Lumber)			
Spandrel Beams (Includes Shoring)			
12" x 48" (4 1/2 BdFt/ SqFt)	SqFt	4.40	1.04
12" x 42" (4 3/4 BdFt/ SqFt)	SqFt	4.45	1.06
12" x 36" (5 BdFt/ SqFt)	SqFt	4.55	1.09
12" x 30" (5 1/4 BdFt/ SqFt)	SqFt	4.70	1.12
8" x 42" (4 3/4 BdFt/ SqFt)	SqFt	4.70	1.05
8" x 36" (5 1/4 BdFt/ SqFt)	SqFt	4.80	1.10
Add for Decks and Safety Rail	SqFt	.90	.35
Interior Beams			
16" x 30" (4 BdFt/ SqFt)	SqFt	4.40	1.02
16" x 24" (4 1/2 BdFt/ SqFt)	SqFt	4.45	1.05
12" x 30" (4 3/4 BdFt/ SqFt)	SqFt	4.50	1.07
12" x 24" (5 1/4 BdFt/ SqFt)	SqFt	4.45	1.14
12" x 16" (6 1/2 BdFt/ SqFt)	SqFt	4.90	1.25
8" x 24" (5 1/4 BdFt/ SqFt)	SqFt	4.80	1.15
8" x 16" (6 1/2 BdFt/ SqFt)	SqFt	5.00	1.25
Add for Beams Carrying Horizontal Shoring	SqFt	.85	.29
Add per Foot for Floor Heights over 10'	SqFt	.42	.10
Add per Floor over 20' Above Grade	SqFt	.12	.09
Add for Splayed Beams	SqFt	1.50	.17
Add for Inverted Beams	SqFt	1.90	.18
Add for Mud Sills (3 Uses)	SqFt	.70	.35

DIVISION #3 - CONCRETE

		UNIT	LABOR	MATL
0302.0	**CONCRETE FORM WORK, Cont'd...**			
.15	Slabs (12' Floor Heights and 3 Uses Lumber)			
.151	Solid Slabs (Includes Shoring) (Metal Adj. Beams or Joists)			
	Horizontal Shoring Method			
	To 5" thick (2' OC x 10' Span) 2.5 BdFt/ SqFt	SqFt	1.75	.64
	5" - 8" thick (2' OC x 15' Span) 2.8 BdFt/ SqFt	SqFt	1.80	.70
	8" - 11" thick (2' OC x 20' Span) 3.1 BdFt/ SqFt	SqFt	2.05	.76
	Add for Heights above 10' per foot	SqFt	.22	.07
	Vertical Shoring Method			
	(Wood or Metal Posts, Purlins & Joists)			
	To 5" thick - 2.7 BdFt/SqFt - 4' x 5'6" OC	SqFt	1.80	.69
	5" - 8" thick - 3.0 BdFt/SqFt - 4' x 5'0" OC	SqFt	2.00	.74
	8" - 11" thick - 3.3 BdFt/SqFt - 4' x 4'6" OC	SqFt	3.40	.78
	Add for Heights above 10' per foot	SqFt	.22	.07
	Add for Adjustable Hardware	SqFt	-	.02
	Add for Cantilevered Slabs	SqFt	1.20	.14
	Add for Drop Panels - No Edge included	SqFt	.22	.10
	Flying Form method (8 Uses)			
	To 5" Thick	SqFt	1.50	.58
	5" - 8" Thick	SqFt	1.55	.59
	8" - 11" Thick	SqFt	1.65	.60
	Add for Heights above 8' per foot	SqFt	.20	.07
	Add for 2 Uses	SqFt	.18	.25
	Add for 4 Uses	SqFt	.14	.18
.152	Pan Slabs - Cost same as solid slabs above plus pan forming costs below.			
	Based on 3 Uses and lease of approximately 4,000 Square Feet			
	20" Pan 8" + 2"	SqFt	.82	.70
	10" + 2"	SqFt	.84	.72
	12" + 2"	SqFt	.87	.74
	14" + 2"	SqFt	.92	.77
	30" Pan 8" + 2 1/2"	SqFt	.79	.67
	10"+2 1/2"	SqFt	.81	.69
	12"+2 1/2"	SqFt	.84	.72
	14"+2 1/2"	SqFt	.87	.74
	19" x 19" Dome Pan - 24" x 24" Joist Center			
	6" + 2"	SqFt	.79	.69
	8" + 2"	SqFt	.82	.71
	10" + 2"	SqFt	.84	.72
	12" + 2"	SqFt	.89	.75
	30" x 30" Dome Pan - 36" x 36" Joist Center			
	10" + 2 1/2"	SqFt	.89	.76
	12" + 2 1/2"	SqFt	.94	.78
	14" + 2 1/2"	SqFt	1.00	.82
	Add for 2 Uses	SqFt	.06	.10
	Add for 1 Use	SqFt	.07	.21
	Deduct for 4 Uses	SqFt	.06	.06
	Deduct for 5 Uses	SqFt	.05	.07
	Pan Slabs: Based on 5 Uses & Lease of 4,000 SqFt			
	40" x 40" Dome Pan - 16" + 3"	SqFt	1.10	1.00
	48" x 48" Joist Center - 18" + 3"	SqFt	1.20	1.20
	Deduct for 6 Uses	SqFt	.06	.09
	Deduct for 7 Uses	SqFt	.05	.10
	Add to above units if subcontracted		30%	10%

DIVISION #3 - CONCRETE

0302.0 CONCRETE FORM WORK, Cont'd...

		UNIT	LABOR	MATL
.16	Stairs and Landing (3 BdFt and 2 Uses)			
	Structural - Contact Area of Soffits, Stringers and Risers	SqFt	3.75	1.17
	On Ground - Contact Area of Stringers and Risers	SqFt	4.90	1.22
.17	Curbs & Platforms (2 BdFt and 2 Uses)	SqFt	3.50	.70
.18	Edge Forms/Bulkheads for Slabs (2 BdFt, 2 Uses)			
	Floor Supported	SqFt	3.45	.70
	Structural	SqFt	4.70	.75
.2	SHORING, SUPPORTS AND ACCESSORIES (All items in this section are included in SqFt costs of Section 0302.1. This section is for cost-keeping & limited analyzing of variables.)			
.21	Horizontal Shoring			
	10' Span - Monthly Rent Cost	Each		1.75
	SqFt Contact Area	SqFt		.39
	15' Span - Monthly Rent Cost	Each		6.50
	SqFt Contact Area	SqFt		.48
	20' Span - Monthly Rent Cost	Each		7.70
	SqFt Contact Area	SqFt		.65
.22	Vertical Shoring (4' OC - 10' High) - Average	SqFt		.28
	Wood Posts with Ellis Shore Hardware			
	Single - W/T Head - New	Each		12.30
	LnFt Beam - 6 Uses	LnFt		.62
	Double - W/T Head & Bracing - New	Each		24.50
	LnFt Beam - 6 Uses	LnFt		1.10
	Metal Posts - Single - Monthly Rent Cost	Each		3.50
	LnFt Beam - 2 Uses per Month	LnFt		.80
	Double - W/T Head & Bracing	LnFt		6.50
	LnFt Beam - 2 Uses per Month	LnFt		1.50
	Tubular Frames - to 10'			
	Mo. Cost - Incl. Head & Base - 10,000#	Each		14.20
	LnFt - Average - 2 Uses per Month	LnFt		1.20
	Deduct for 6,000#	Each		25%
.23	Column Clamps and Steel Strapping			
	Clamps - Monthly Rent Cost - Steel	Each		4.30
	Lock Fast	Each		8.00
	Gang Form	Each		7.00
	Contact Area - 2 Uses per Month	SqFt.		35
	Strapping - 3/4" x .025 (1.10 lb)	LnFt		.25
	3/4" x .023 (1.00 lb)	LnFt		.24
	Contact Area of Forms	SqFt		.15
.24	Wall and Beam Brackets			
	Brackets (2' OC) - Monthly Rent Cost	Each		.60
	New Cost	Each		5.80
	Contact Area - 4 Uses per Month	SqFt		.14
.25	Accessories - Beam Hangers			
	Tyloops, Lags, Washers & Bolts - Heavy	LnFt		.90
	Light	LnFt		.60

DIVISION #3 - CONCRETE

0302.0 CONCRETE FORM WORK, Cont'd...

.3	SPECIALTIES	UNIT	LABOR	MATERIAL
.31	Nails, Wire and Ties			
	Nails 6 d common (50# box)	CWT	-	44.00
	8 d common	CWT	-	46.00
	16 d common	CWT	-	50.00
	8 d box	CWT	-	74.00
	16 d box	CWT	-	78.00
	9 ga concrete nails (3/4" to 3")	CWT	-	130.00
	Add for Coated Nails	CWT	-	60%
	Wire - Black Annealed 9 ga (100# Coil)	CWT	-	76.00
	16 ga (100# Coil)	CWT	-	73.00
	Average Nail and Wire Cost for Formwork	BdFt	-	.02
	or	SqFt	-	.03
	Ties - Breakback -8" 3000# 5" ends	Each	-	.62
	12" 3000# 5" ends	Each	-	.70
	16" 3000# 5" ends	Each	-	.75
	Add for 5000#	Each	-	.13
	Add for 8" ends	Each	-	.07
	Average Tie Cost for Formwork (2' OC)			
	Wall Area	SqFt	-	.34
	Contact Area Form	SqFt	-	.18
	Gang Form Contact Area	SqFt	-	.19
	Coil Bolts - 8"	Each	-	1.25
	12"	Each	-	1.40
.32	Anchors and Inserts (including Layout)			
	Anchor Bolts (Concrete) -			
	1/2" x 12"	Each	1.00	.70
	5/8" x 12"	Each	1.10	1.80
	3/4" x 12"	Each	1.25	2.65
	Ceiling Type Inserts			
	Adjustable - 1/2" Bolt Size x 3'	Each	.80	2.10
	5/8" Bolt Size x 3'	Each	.85	2.15
	3/4" Bolt Size x 3'	Each	.90	2.50
	Continuous Slotted	Each	.70	1.65
	Threaded - 1/2" Bolt Size	Each	.80	3.50
	5/8" Bolt Size	Each	.85	5.70
	3/4" Bolt Size	Each	.90	6.75
	Hanger Wire Type - Drive In	Each	.70	.17
	Shell	Each	.70	.21
	Ferrule Type - 1/2"	Each	.75	.98
	5/8"	Each	.80	1.20
	3/4"	Each	.85	1.70
	T-Hanger - 4"	Each	.75	.90
	10"	Each	.75	1.35
	14"	Each	.90	1.80
	Wall and Beam Type Insert			
	Dovetail Slot - 24 ga (No Anchors)	LnFt	.30	38
	22 ga (No Anchors)	LnFt	.37	.75
	Flashing Reglets	LnFt	.50	.60
	Shelf Angle Inserts - Add for Bolts,			
	Washers, and Nuts -5/8"	Each	.80	2.70
	3/4"	Each	.85	3.40

DIVISION #3 - CONCRETE

0302.0	CONCRETE FORM WORK, Cont'd...	UNIT	LABOR	MATERIAL
	Bolts - 5/8" x 2"	Each	.70	1.10
	3/4" x 2"	Each	.71	1.70
	3/4" x 3"	Each	.73	2.35
	Nuts - 5/8"	Each	.26	.20
	3/4"	Each	.29	.30
	Washers	Each	.07	.23
	Threaded Rod - 3/8" x 36" (Galv)	Each	.90	1.20
	1/2" x 36" (Galv)	Each	1.00	2.10
	5/8" x 36" (Galv)	Each	1.20	3.25
.33	Chamfers, Drips and Reveals (1 Use)			
	Chamfers - 3/4" Metal	LnFt	.24	.45
	Plastic with Tail	LnFt	.26	.50
	without Tail	LnFt	.26	.40
	Wood	LnFt	.23	.15
	Reveals - 3/4" Plastic	LnFt	.34	1.20
	Drips	LnFt	.30	.40
.34	Stair Nosings and Treads			
	Nosings, Curb Bar - Steel - Galvanized	LnFt	1.00	4.25
	Treads 3" x 1/2" - Cast Iron	LnFt	1.25	7.00
	Extruded Aluminum	LnFt	1.25	6.50
	Cast Aluminum	LnFt	1.25	8.00
.35	Water Stops			
	Polyvinyl Chloride - 6" x 3/16"	LnFt	.72	1.25
	6" Centerbulb	LnFt	.76	2.40
	9" x 3/16"	LnFt	.83	1.90
	9" Centerbulb	LnFt	.90	3.00
	Add for 3/8"	LnFt	.15	50%
	Rubber (Neoprene) - 6" Centerbulb	LnFt	.80	7.80
	9" Centerbulb	LnFt	.90	15.00
	Add per Splice	Each	6.75	-
.36	Divider Strips - White Metal, 16 ga x 1 1/4"	LnFt	.61	.85
	1/2" x 1 1/4"	LnFt	.62	.90
	3/16" x 1 1/4"	LnFt	.62	.90
	1/4" x 1 1/4"	LnFt	.62	1.50
	Add for Brass 1/8"	LnFt	.10	.80
.37	Tongue and Groove Joint Forms			
	Asphalt - 3 1/2" x 1/8"	LnFt	.66	.70
	5 1/2" x 1/8"	LnFt	.73	.85
	Metal - 3 1/2"	LnFt	.67	.58
	5 1/2"	LnFt	.73	.75
	Wood - 3 1/2"	LnFt	.67	.22
	5 1/2"	LnFt	.73	.34
	Plastic - 3 1/2"	LnFt	.67	.80
	Add for Stakes	Each	.65	.50
.38	Bearing Pads and Shims			
	Vinyl - 1/8" x 6" x 24"	SqFt	.75	1.90
	1/4" x 6" x 24"	SqFt	.86	3.80
	Neoprene - 1/8" x 36" (70 Duro)	SqFt	.65	4.90
	1/4" x 36"	SqFt	.70	8.75
.4	EQUIPMENT			

See Pages 1 - 6A and 1 - 7A. Average 5% Labor Cost.

Add to Labor Formwork Costs Above for Winter Construction 10% to 25%

DIVISION #3 - CONCRETE

0303.0 REINFORCING STEEL (S&L) (Ironworkers)

	UNIT	LABOR Subcont	MATERIAL Plain	Epoxy
.1 BARS (Based on Local Trucking)				
1/4" to 5/8" Large Size Job (100 Tons Up)	Ton	425	610	1,400
Medium Job (20 - 99 Tons Up)	Ton	475	650	1,460
Small Size Job (5 - 19 Tons)	Ton	500	700	1,570
3/4" to 1 1/4" Large Job (100 Tons Up)	Ton	440	610	1,380
Medium Job (10 - 99 Tons)	Ton	450	620	1,400
Small Size Job (5 - 19 Tons)	Ton	470	660	1,480
1 1/2" to 2" Average	Ton	460	640	1,440
Add For Galvanizing	Ton	55	440	-
Add for Bending - Light #2 to #5	Ton	-	240	-
Heavy #6 and Larger	Ton	-	120	-
Add for Size Extras - Light #2 to #4	Ton	-	145	-
Heavy #3 and Larger	Ton	-	110	-
Add for Spirals - Shop Assembled - Hot Rolled	Ton	-	165	-
Cold Rolled	Ton	-	230	-
Unassembled - Hot Rolled	Ton	-	140	-
Cold Rolled	Ton	-	220	-
Add for Splicing: #11 Bars - Buttweld	Each	30.00	3.00	-
#14 Bars - Buttweld	Each	40.00	4.30	-
#11 Bars - Mech. Buttweld	Each	42.00	11.00	-
#14 Bars - Mech. Buttweld	Each	48.00	12.00	-

	UNIT	LABOR	MATERIAL
BARS (Based on Local Trucking)			
Tables of Weights & Prices by LnFt -	Ton	-	800.00
Small job from warehouse stock for local delivery	Lb	-	.40
# 2 Bar 1/4" or .250 .167 lb	LnFt	.12	.11
# 3 Bar 3/8" or .375 .376 lb	LnFt	.15	.15
# 4 Bar 1/2" or .500 .668 lb	LnFt	.20	.21
# 5 Bar 5/8" or .625 1.043 lb	LnFt	.29	.33
# 6 Bar 3/4" or .750 1.502 lb	LnFt	.40	.45
# 7 Bar 7/8" or .875 2.044 lb			
# 8 Bar 1" or 1.000 2.670 lb			
# 9 Bar 1 1/8" or 1.128 3.400 lb			
#10 Bar 1 1/4" or 1.270 4.303 lb			
#11 Bar 1 3/8" or 1.410 5.313 lb			
#14 Bar 1 1/2" or 1.693 7.650 lb			
#18 Bar 2 1/4" or 2.257 13.600 lb			

.2 ACCESSORIES
Additional sizes, but below are close to standard and are included in tonnage costs above

	UNIT	LABOR	MATERIAL
1/4" to 1 1/2" Slab Bolsters	LnFt	-	.30
1/2" to 2" Beam Bolsters - Upper & Lower	LnFt	-	.58
2" to 3" Beam Bolsters - Upper & Lower	LnFt	-	.90
3" High Chairs - Continuous	LnFt	-	.45
3/4" x 4" to 6" Joist Chairs	Each	-	.40
1/2" Dowel Bar Tubes - Plastic and Metal	Each	.45	.28
Add for Galvanized Accessories	-	-	.15
Add for Plastic Accessories	-	-	.18

.3 POST TENSION STEEL - PRESTRESSED IN FIELD

	UNIT	LABOR	MATERIAL
Ungrouted - 100 kips	Lb	.72	1.25
200 kips	Lb	.67	1.30
Deduct for Added 100 kips	Lb	.02	-
Add for Chairs	Lb	-	.12

DIVISION #3 - CONCRETE

0303.0 REINFORCING STEEL, Cont'd...

	UNIT	LABOR	MATERIAL
.4 WIRE FABRIC (750 SqFt Roll or 120 SqFt Sheet)			
6" x 6" x 10/10 (W 1.4)	SqFt	.12	.09
6" x 6" x 8/8 (W 2.1)	SqFt	.14	.13
6" x 6" x 6/6 (W 2.9)	SqFt	.17	.14
6" x 6" x 4/4 (W 4.0)	SqFt	.20	.17
4" x 4" x 4/4 (W 4.0)	SqFt	.25	.24
4" x 6" x 8/8 (W 2.0)	SqFt	.17	.15
Add for Epoxy Coated (120 SqFt Sheet)	SqFt	.02	.20

0304.0 SPECIALTY PLACED CONCRETE (No Fees)

	UNIT		MATERIAL
.1 LIFT SLAB CONSTRUCTION			
Concrete - 6" Slab	SqFt		3.80
Formwork	SqFt		1.10
Reinforcing Steel	SqFt		1.25
Lifting	SqFt		1.25
Brace and Assemble	SqFt		.80
		TOTAL	8.20
.2 POST TENSIONED OR STRESSED IN PLACE CONCRETE			
Concrete - 6" Slab and Columns	SqFt		4.25
Formwork (Assume Flying Forms)	SqFt		4.50
Post Tensioning and Reinforcing Steel	SqFt		3.00
		TOTAL	11.75
.3 TILT UP CONSTRUCTION (Incl. Concrete Columns)			
Concrete - 6" Slab	SqFt		4.00
Formwork	SqFt		2.65
Reinforcing Steel	SqFt		2.55
Tilting Up and Bracing	SqFt		1.25
		TOTAL	10.45

	UNIT	1"	2"	3"
.4 PNEUMATICALLY PLACED CONCRETE (Gunite)				
Walls - 1" Layers with Mesh	SqFt	3.00	4.75	5.70
Columns - 1" Layers with Mesh	SqFt	5.80	6.20	7.20
Roofs	SqFt	3.80	4.40	5.00

0305.0 PRECAST PRESTRESSED CONCRETE (L&M)

	Spans to Approximate	UNIT	COST
.1 BEAMS AND COLUMNS			
Beams	20'	LnFt	85.00
	30'	LnFt	102.00
Columns - 12" x 12" x 12' with Plates		LnFt	110.00
16" x 16" x 12' with Plates		LnFt	135.00
.2 DECKS (Flat) - 24", 40", 48", 72" and 96" Wide - 50 lb. Roof Load			
4"	12'	SqFt	4.60
6"	18'	SqFt	5.10
8"	24'	SqFt	5.50
10"	30'	SqFt	6.00
12"	50'	SqFt	7.20
.3 DOUBLE T (Flat)			
8' x 20" - 30" Deep - 60 lb. Roof Load	40'	SqFt	5.90
	50'	SqFt	6.20
	60'	SqFt	7.20
	70'	SqFt	8.00
	80'	SqFt	8.60
SINGLE T			
8' x 36" - 48" Deep	60'	SqFt	9.40
	80'	SqFt	10.40
	100'	SqFt	11.60
	120'	SqFt	13.00
Add for Floor Decks (100# Load)		SqFt	10%
Add for Sloped Installation to Decks & T's		SqFt	10%
Add for 2" Topping for Floors - If Required		SqFt	3.00
Add per Mobilization		Each	2,700.00

DIVISION #3 - CONCRETE

		UNIT	COST
0305.0	**PRECAST PRESTRESSED CONCRETE, Cont'd...**		
.4	WALLS		
	Add to Flat Decks 0305.2 Above	SqFt	1.65
	Add to Double T 0305.3 Above	SqFt	1.80
	Add for Core Wall Type	SqFt	1.60
	Add for Insulated Type	SqFt	1.70
0306.0	**PRECAST CONCRETE (M)**		
.1	BEAMS AND COLUMNS- Same as 0305.1		
.2	DECKS- Same as 0305.2		
.3	PANELS OR FACING UNITS (L&M)		
	Insulated - 8' x 16' x 12"		
	Exposed Aggregate		
	Gravel	SqFt	13.00
	Quartz & Marble	SqFt	15.00
	Trowel or Brush		
	Grey Cement	SqFt	12.50
	White Cement	SqFt	13.75
	Deduct for Non-Insulated 8' x 16' x 8"	SqFt	1.15
	Add for Banded	SqFt	.45
.4	PLANK (S)		
	2" x 24" 6' Span	SqFt	4.25
	3" x 24" 8' Span	SqFt	4.75

			LABOR	MATERIAL
.5	SPECIALTIES (S)			
	Curbs 6" x 10" x 8'	LnFt	2.05	6.20
	8" x 10" x 8'	LnFt	2.45	6.75
	Copings 5" x 13"	LnFt	5.50	9.00
	Sills & Stools 5" x 6"	LnFt	4.40	11.20
	Splash Blocks 3" x 36" x 16"	Each	9.00	22.30
	5" x 36" x 16"	Each	11.30	23.30
	8" x 36" x 16"	Each	15.00	27.70
	Medians	LnFt	8.40	36.00

				COST
0307.0	**PRECAST CEMENTITIOUS PLANK (L&M)**			
.1	EXPANDED MINERALS (Includes Clips, Rods & Grout)			
		SPANS TO APPROX.		
	3" x 24"	6'-0"	SqFt	3.25
	4" x 24"	9'-6"	SqFt	3.50
.2	GYPSUM (Metal Edged)			
	2" x 15"	4'-0"	SqFt	3.35
	2" x 15"	7'-0"	SqFt	3.60
.3	WOOD FIBRE			
	2" x 32"	3'-0"	SqFt	3.10
	2 1/2" x 32"	3'-6"	SqFt	3.20
	3" x 32"	4'-0"	SqFt	3.40
0308.0	**POURED IN PLACE CEMENTITIOUS CONCRETE (L&M)**			
.1	EXPANDED MINERALS - 2"		SqFt	2.05
.2	EXPANDED MINERALS & ASPHALT - 2"		SqFt	2.25
.3	GYPCRETE - 1/4"		SqFt	1.05
	1"		SqFt	1.35

DIVISION #3 - CONCRETE - QUICK ESTIMATING

All examples below are representative samples and have great variables because of design efficiencies, size, spans, weather, etc. All figures include a contractor's fee of 10% for Footings and Slabs on Ground (low risk) and a 15% fee for Walls, Columns, Beams and Slabs (high risk). All totals include material prices as follows: Concrete $78.00 CuYd, Lumber $500.00 M.B.F. (3 uses), Plywood $1.00 SqFt, and Reinforcing Steel $650.00 Ton. Taxes and insurance on Labor added at 35%. Includes General Conditions at 10%. CuYd Costs are for information only--not quick estimating.

	Unit	Unit Cost With ReinSteel	Unit Cost Without ReinSteel	CuYd Cost With ReinSteel
FOOTINGS (Including Hand Excavation)				
Continuous 24" x 12" (3 # 4 Rods)	LnFt	19.10	17.30	258.00
36" x 12" (4 # 4 Rods)	LnFt	24.70	24.50	223.00
20" x 10" (2 # 5 Rods)	LnFt	16.20	15.50	316.00
16" x 8" (2 # 4 Rods)	LnFt	13.85	16.10	415.00
Pad 24" x 24" x 12" (4 # 5 E.W.)	Each	67.00	50.00	453.00
36" x 36" x 14" (6 # 5 E.W.)	Each	135.00	95.90	405.00
48" x 48" x 16" (8 # 5 E.W.)	Each	225.00	185.00	380.00
WALLS (#5 Rods 12" O.C. - 1 Face)				
8" Wall (# 5 12" O.C. E.W.)	SqFt	13.40	11.90	542.00
12" Wall (# 5 12" O.C. E.W.)	SqFt	14.60	13.10	394.00
16" Wall (# 5 12" O.C. E.W.)	SqFt	16.30	14.90	330.00
Add for Steel - 2 Faces - 12" Wall	SqFt	1.75	-	46.00
Add for Pilastered Wall - 24" O.C.	SqFt	.80	-	21.50
Add for Retaining or Battered Type	SqFt	2.45	-	66.00
Add for Curved Walls	SqFt	4.50	-	131.00
COLUMNS				
Sq Cornered 8" x 8" (4 # 8 Rods)	LnFt	32.50	23.40	1,908.00
12" x 12" (6 # 8 Rods)	LnFt	53.50	43.30	1,445.00
16" x 16" (6 # 10 Rods)	LnFt	73.60	54.70	1,105.00
20" x 20" (8 # 20 Rods)	LnFt	96.40	70.00	935.00
24" x 24" (10 # 11 Rods)	LnFt	125.00	84.10	844.00
Round 8" (4 # 8 Rods)	LnFt	23.40	15.90	1,675.00
12" (6 # 8 Rods)	LnFt	36.55	22.40	1,190.00
16" (6 # 10 Rods)	LnFt	49.50	31.40	940.00
20" (8 # 20 Rods)	LnFt	70.20	46.80	865.00
24" (10 # 11 Rods)	LnFt	103.00	64.50	845.00
BEAMS				
Spandrel 12" x 48" (33 # Rein. Steel)	LnFt	125.40	98.70	845.00
12" x 42" (26 # Rein. Steel)	LnFt	110.20	85.60	847.00
12" x 36" (21 # Rein. Steel)	LnFt	89.70	74.00	807.00
12" x 30" (15 # Rein. Steel)	LnFt	80.50	72.80	870.00
8" x 48" (26 # Rein. Steel)	LnFt	103.90	96.10	1,035.00
8" x 42" (21 # Rein. Steel)	LnFt	99.50	87.10	1,115.00
8" x 36" (16 # Rein. Steel)	LnFt	91.40	76.20	1,235.00
Interior 16" x 30" (24 # Rein. Steel)	LnFt	72.80	65.80	587.00
16" x 24" (20 # Rein. Steel)	LnFt	71.40	61.60	715.00
12" x 30" (17 # Rein. Steel)	LnFt	78.20	61.60	725.00
12" x 24" (14 # Rein. Steel)	LnFt	59.60	52.80	805.00
12" x 16" (12 # Rein. Steel)	LnFt	49.80	40.90	995.00
8" x 24" (13 # Rein. Steel)	LnFt	58.70	49.70	1,174.00
8" x 16" (10 # Rein. Steel)	LnFt	44.30	38.60	1,330.00

DIVISION #3 - CONCRETE - QUICK ESTIMATING

	Unit	Unit Cost With ReinSteel	Unit Cost Without ReinSteel	CuYd Cost With ReinSteel
SLABS				
Solid -				
4" Thick	SqFt	8.55	6.85	692.00
5" Thick	SqFt	9.90	7.60	638.00
6" Thick	SqFt	10.85	7.30	586.00
7" Thick	SqFt	12.25	9.80	563.00
8" Thick	SqFt	13.80	11.70	559.00
Deduct for Post Tensioned Slabs	SqFt	.80	-	-
Pan -				
Joist -20" Pan - 10" x 2"	SqFt	11.90	10.40	-
12" x 2"	SqFt	13.40	11.40	-
30" Pan - 10" x 2 1/2"	SqFt	11.90	10.15	-
12" x 2 1/2"	SqFt	13.40	11.25	-
Dome -19" x 19" - 10" x 2"	SqFt	12.90	10.10	-
12" x 2"	SqFt	13.60	10.20	-
30" x 30" - 10" x 2 1/2"	SqFt	11.70	10.00	-
12" x 2 1/2"	SqFt	12.75	10.10	-
COMBINED COLUMNS, BEAMS AND SLABS				
20' Span	SqFt	17.50	-	-
30' Span	SqFt	18.50	-	-
40' Span	SqFt	19.70	-	-
50' Span	SqFt	21.30	-	-
60' Span	SqFt	25.40	-	-
STAIRS (Including Landing)				
4' Wide 10' Floor Heights 16 Risers	Riser	156.00	-	1,270.00
5' Wide 10' Floor Heights 16 Risers	Riser	189.00	-	1,350.00
6' Wide 10' Floor Heights 16 Risers	Riser	211.00	-	1,300.00

	Unit	Unit Cost With Mesh	CuYd Cost With Mesh
SLABS ON GROUND			
4" Concrete Slab			
(6 6/10 - 10 Mesh - 5 1/2 Sack Concrete,			
Trowel Finished, Cured & Truck Chuted)	SqFt	3.50	267.00
Add per Inch of Concrete	SqFt	.55	
Add per Sack of Cement	SqFt	.13	
Deduct for Float Finish	SqFt	.08	
Deduct for Brush or Broom Finish	SqFt	.05	
Add for Runway and Buggied Concrete	SqFt	.16	
Add for Vapor Barrier (4 mil)	SqFt	.14	
Add for Sub - Floor Fill (4" sand/gravel)	SqFt	.40	
Add for Change to 6 6/8 - 8 Mesh	SqFt	.07	
Add for Change to 6 6/6 - 6 Mesh	SqFt	.13	
Add for Sloped Slab	SqFt	.20	
Add for Edge Strip (sidewalk area)	SqFt	.20	
Add for ½" Expansion Joint (20' O.C.)	SqFt	.07	
Add for Control Joints (keyed/ dep.)	SqFt	.15	
Add for Control Joints (joint filled)	SqFt	.16	
Add for Control Joints - Saw Cut (20' O.C.)	SqFt	.27	
Add for Floor Hardener (1 coat)	SqFt	.12	
Add for Exposed Aggregate - Washed Added	SqFt	.38	
Retarding Added	SqFt	.35	
Seeding Added	SqFt	.40	
Add for Light Weight Aggregates	SqFt	.40	
Add for Heavy Weight Aggregates	SqFt	.28	
Add for Winter Production Loss/Cost	SqFt	.33	

See 0206.2 for Curbs, Gutters, Walks and other Exterior Concrete
See 0302.53 for Topping and Leveling of Floors

DIVISION #3 - CONCRETE - QUICK ESTIMATING

	UNIT	COST
TOPPING SLABS		
2" Concrete	SqFt	2.55
3" Concrete	SqFt	3.15
No Mesh or Hoisting		
PADS & PLATFORMS (Including Form Work & Reinforcing)		
4"	SqFt	6.20
6"	SqFt	7.65
PRECAST CONCRETE ITEMS		
Curbs 6" x 10" x 8"	LnFt	10.00
Sills & Stools 6"	LnFt	23.00
Splash Blocks 3" x 16"	Each	55.00
MISCELLANEOUS ADDITIONS TO ABOVE CONCRETE IF NEEDED		
Abrasives - Carborundum - Grits	SqFt	.75
Strips	LnFt	2.60
Bushammer - Green Concrete	SqFt	1.65
Cured Concrete	SqFt	2.20
Champers - Plastic 3/4"	LnFt	.85
Wood 3/4"	LnFt	.55
Metal 3/4"	LnFt	.82
Colors Dust On	SqFt	.65
Integral (Top 1")	SqFt	1.20
Control Joints - Asphalt 1/2" x 4"	LnFt	.75
1/2" x 6"	LnFt	.85
PolyFoam 1/2" x 4"	LnFt	1.00
Dovetail Slots 22 Ga	LnFt	1.30
24 Ga	LnFt	.85
Hardeners Acrylic and Urethane	SqFt	.22
Epoxy	SqFt	.30
Joint Sealers Epoxy	LnFt	2.40
Rubber Asphalt	LnFt	1.30
Moisture Proofing - Polyethylene 4 mil	SqFt	.15
6 mil	SqFt	.18
Non-Shrink Grouts - Non - Metallic	CuFt	35.00
Aluminum Oxide	CuFt	25.00
Iron Oxide	CuFt	38.00
Reglets Flashing	LnFt	1.40
Sand Blast - Light	SqFt	1.20
Heavy	SqFt	2.20
Shelf Angle Inserts 5/8"	LnFt	4.20
Stair Nosings Steel - Galvanized	LnFt	6.10
Tongue & Groove Joint Forms - Asphalt 3 1/2"	LnFt	1.75
Wood 3 1/2"	LnFt	1.25
Metal 3 1/2"	LnFt	1.65
Treads - Extruded Aluminum	LnFt	8.50
Cast Iron	LnFt	9.50
Water Stops Center Bulb - Rubber - 6"	LnFt	9.80
9"	LnFt	17.00
Polyethylene - 6"	LnFt	3.80
9"	LnFt	4.50

DIVISION #4 - MASONRY

Wage Rates (Including Fringes) & Location Modifiers
July 2002-2003

	Metropolitan Area		Bricklayer		Tender	Wage Rate Location Modifier
1.	Akron	*	35.53	*	28.17	115
2.	Albany-Schenectady-Troy		33.40		26.51	116
3.	Atlanta		21.55		14.23	74
4.	Austin	**	22.93	**	16.16	74
5.	Baltimore		25.94		18.56	90
6.	Birmingham	**	23.69	**	18.05	72
7.	Boston		43.22		31.75	145
8.	Buffalo-Niagara Falls		37.73		32.13	127
9.	Charlotte	**	20.56	**	14.15	66
10.	Chicago-Gary	*	39.51	*	34.28	126
11.	Cincinnati		29.44		24.92	94
12.	Cleveland		34.56		30.51	113
13.	Columbus	*	31.44	*	23.76	105
14.	Dallas-Fort Worth	**	21.04	**	14.71	69
15.	Dayton	*	31.58	*	24.46	101
16.	Denver-Boulder	*	25.91	*	16.60	83
17.	Detroit	*	39.83	*	30.20	126
18.	Flint	*	33.86	*	25.56	110
19.	Grand Rapids	*	34.00	*	23.25	110
20.	Greensboro-West Salem	**	20.72	**	14.18	66
21.	Hartford-New Britain	*	36.11	*	28.85	118
22.	Houston	**	23.54	**	17.15	82
23.	Indianapolis		30.98		24.30	101
24.	Jacksonville	**	24.86	**	16.26	75
25.	Kansas City		32.34		24.48	97
26.	Los Angeles-Long Beach	*	36.47	*	30.51	128
27.	Louisville	**	23.18	**	19.59	80
28.	Memphis	**	24.85	**	13.67	88
29.	Miami	**	24.87	**	17.12	80
30.	Milwaukee	*	35.25	*	29.87	115
31.	Minneapolis-St. Paul		35.84		30.77	112
32.	Nashville	**	23.60	**	15.88	78
33.	New Orleans	**	22.07	**	14.80	74
34.	New York	*	52.07	*	42.61	177
35.	Norfolk-Portsmouth	**	23.32	**	15.88	76
36.	Oklahoma City	**	24.73	**	16.82	88
37.	Omaha-Council Bluffs	**	25.64	**	20.31	82
38.	Orlando	**	25.89	**	19.32	84
39.	Philadelphia	*	39.16	*	32.83	127
40.	Phoenix	**	29.07	**	18.30	94
41.	Pittsburgh		34.30		24.96	111
42.	Portland	*	36.35	*	29.00	116
43.	Providence-Pawtucket	*	33.90	*	27.93	110
44.	Richmond	**	23.54	**	15.77	76
45.	Rochester	*	35.15	*	27.31	114
46.	Sacramento	*	43.27		31.27	138
47.	St. Louis		35.07		30.57	112
48.	Salt Lake City-Ogden	**	21.91	**	17.54	68
49.	San Antonio	**	21.31	**	12.92	67
50.	San Diego	*	33.18	*	31.12	127
51.	San Francisco-Oakland-San Jose		43.27		31.29	137
52.	Seattle-Everett		36.22		30.99	112
53.	Springfield-Holyoke-Chicopee		33.61		32.44	110
54.	Syracuse	*	31.17	*	25.11	99
55.	Tampa-St. Petersburg	**	25.90	**	18.00	85
56.	Toledo	*	32.98	*	26.34	107
57.	Tulsa	**	25.01	**	16.79	88
58.	Tucson	**	29.28	**	21.04	95
59.	Washington D.C.	*	26.70	*	17.82	87
60.	Youngstown-Warren	*	32.40	*	27.99	106
	AVERAGE		**30.58**		**23.46**	

Note: Labor Rates Increased by .15 for Tending (average)
Impact Ratio: Labor 65%, Material 35%

* Contract Not Settled - Wage Interpolated
** Non Signatory or Open Shop Rate

DIVISION #4 - MASONRY

			PAGE
0401.0	**BRICK MASONRY (04210)**		**4-4**
.1	FACE BRICK		4-4
.11		Conventional	4-4
.12		Econo	4-4
.13		Panel	4-4
.14		Norman	4-4
.15		King	4-4
.16		Norwegian	4-4
.17		Saxon - Jumbo	4-4
.18		Adobe - Mexican	4-4
.2	COATED(Ceramic Veneer) (04250)		4.4
.3	COMMON BRICK		4-4
.4	FIRE BRICK (04550)		4-4
0402.0	**CONCRETE BLOCK (04220)**		**4-5**
.1	CONVENTIONAL		4-5
.11		Foundation	4-5
.12		Backup	4-5
.13		Partitions	4-5
.2	ORNAMENTAL OR SPECIAL EFFECT		4-6
.21		Shadowall and Hi-Lite	4-6
.22		Scored	4-6
.23		Breakoff	4-6
.24		Split Face	4-6
.25		Rock Faced	4-6
.26		Adobe (Slump)	4-6
.3	SCREEN WALL		4-6
.4	INTERLOCKING		4-6
.5	SOUND		4-6
.6	BURNISHED		4-7
.7	PREFACED UNITS		4-7
0403.0	**CLAY BACKING AND PARTITION TILE (04240)**		**4-7**
0404.0	**CLAY FACING TILE (GLAZED) (04245)**		**4-8**
.1	6 T SERIES		4-8
.2	8 W SERIES		4-8
0405.0	**GLASS UNITS (04270)**		**4-8**
0406.0	**TERRA COTTA (04280)**		**4-8**
0407.0	**MISCELLANEOUS UNITS**		**4-8**
.1	COPING TILE		4-8
.2	FLUE LINING (04452)		4-8
0408.0	**NATURAL STONE (04400)**		**4-9**
.1	CUT STONE (04420)		4-9
.11		Limestone	4-9
.12		Marble	4-9
.13		Granite	4-9
.14		Slate	4-9
.2	ASHLAR STONE		4-9
.21		Limestone	4-9
.22		Marble	4-9
.23		Granite	4-9
.3	ROUGH STONE (04410)		4-9
.31		Flagstone and Rubblestone (04400)	4-9
.32		Field Stone or Boulders	4-9
.33		Light Weight Boulders (Igneous)	4-9
.34		Light Weight Limestone	

DIVISION #4 - MASONRY

		PAGE
0409.0	**PRECAST VENEERS AND SIMULATED MASONRY (04430)**	**4-9**
.1	ARCHITECTURAL PRECAST STONE (04435)	4-9
.2	PRECAST CONCRETE	4-9
.3	MOSAIC GRANITE PANELS	4-9
0410.0	**MORTARS (04100)**	**4-10**
.1	PORTLAND CEMENT AND LIME	4-10
.2	MASONRY CEMENT	4-10
.3	EPOXY CEMENT	4-10
.4	HIGH EARLY CEMENT	4-10
.5	FIRE RESISTANT	4-10
.6	ADMIXTURES	4-10
.7	COLORING	4-10
0411.0	**CORE FILLING FOR REINFORCED CONCRETE MASONRY (04290)**	**4-10**
0412.0	**CORE AND CAVITY FILLING FOR INSULATED MASONRY**	**4-11**
.1	LOOSE	4-11
.2	RIGID	4-11
0413.0	**ACCESSORIES AND SPECIALTIES (04150)**	**4-12**
.1	ANCHORS (04170)	4-12
.2	TIES	4-12
.3	REINFORCEMENT (04160)	4-12
.4	FIREPLACE AND CHIMNEY ACCESSORIES	4-12
.5	CONTROL AND EXPANSION JOINTS (04180)	4-13
.6	FLASHINGS	4-13
0414.0	**CLEANING AND POINTING (04510)**	**4-13**
.1	BRICK AND STONE	4-13
.2	BLOCK	4-13
.3	FACING AND TILE	4-13
0415.0	**MASONRY RESTORATION (04500)**	**4-13**
.1	RAKING FILLING AND TUCKPOINTING	4-13
.2	SANDBLASTING	4-13
.3	STEAM CLEANING	4-13
.4	HIGH PRESSURE WATER	4-13
.5	BRICK REPLACEMENT	4-13
.6	WATERPROOFING	4-13
0416.0	**INSTALLATION OF STEEL AND MISCELLANEOUS EMBEDDED ITEMS**	**4-13**
0417.0	**EQUIPMENT AND SCAFFOLD**	**4-13**
.1	EQUIPMENT	4-13
.2	SCAFFOLD	4-13
4A	**QUICK ESTIMATING SECTIONS**	4A-14, 4A-15, 4A-16, 4A-17, 4A-18
4B	**TABLES FOR:** Mortar Quantities	4B-19, 4B-20
	UNITS PER SQUARE FOOT	
	Placed Per Man Day	
	HAULING AND UNLOADING COSTS	
	WEIGHTS PER UNIT	

The unit costs in this Division are priced as being done by masonry or General Contractors. See Section 4A for Total Subcontracted Cost. In general, the pricing is based on Modular Units, and materials F.O.B. job site in truckload lots. The basic labor units do include mortar mixing, low scaffolding, tending, and labor fringe benefits. Cleaning, high scaffold, equipment reinforcements, taxes and insurance, etc. must be added to basic labor and material units. The basis for Labor Unit Costs can be found as follows: as to labor rate mean, Page 4-1, and as to number of units placed per day, Table 4B. Unit Costs are based on an average crew of 4 Bricklayers to 3 Tenders (including Mortar Mixer and Operators) for Brick and Stone Work, and 1 Bricklayer to 1 Tender for Block and Tile Work.

DIVISION #4 - MASONRY

0401.0 BRICK MASONRY (Bricklayers)   UNIT   LABOR   MATERIAL
  .1 FACE BRICK (See Table 4B for Mfg. Size)   (Truckload)

		Nominal Dimensions	UNIT	LABOR	MATERIAL
.11	Conventional	8" x 2-2/3" x 4"			
	Running Bond		M pcs	700.00	490.00
	Common Bond (6 Cs. Headers)		M pcs	710.00	490.00
	Stack Bond		M pcs	730.00	490.00
	Dutch & English Bond		M pcs	770.00	490.00
	Flemish		M pcs	770.00	490.00
.12	Economy	8" x 4" x 4"	M pcs	845.00	720.00
.13	Panel (Triple)	8" x 8" x 4"	M pcs	1,380.00	1,600.00
		8" x 16" x 4"	M pcs	2,270.00	3,200.00
.14	Norman	12" x 2-2/3" x 4"	M pcs	830.00	720.00
.15	King	7-5/8", 8-5/8" & 9-5/8" x 2-5/8" x 4"	M pcs	760.00	490.00
.16	Norwegian	12" x 3-1/5" x 4"	M pcs	900.00	820.00
.17	Saxon-Utility 3" wall	12" x 4" x 3" (thin)	M pcs	1,090.00	980.00
	4" wall	12" x 4" x 4"	M pcs	1,100.00	1,010.00
	6" wall	12" x 4" x 6"	M pcs	1,330.00	1,430.00
	8" wall	16" x 3-5/8" x 3-5/8"	M pcs	1,510.00	1,840.00
.18	Meridian	12" x 2-1/4" x 4"	M pcs	690.00	1500.00
.19	Adobe Brick - Mexican	8" x 2-3/8" x 4"	M pcs	830.00	800.00
		12" x 2-3/8" x 4"	M pcs	960.00	700.00
.2	COATED BRICK (Ceramic Veneer)		M pcs	785.00	1,100.00
.3	COMMON BRICK - Clay		M pcs	560.00	300.00
	Concrete		M pcs	560.00	270.00
.4	FIRE BRICK - Light Duty (domestic)		M pcs	700.00	1,100.00
	Medium Duty		M pcs	710.00	1,750.00
	Heavy Duty		M pcs	730.00	2,300.00
	Add for Fireplaces		M pcs	550.00	-
Add: Less than Truckload Lot (14M pcs)					15%
Add: Full Size Brick			M pcs	4%	8%
Add: Ceramic Coating - to Common Brick			M pcs	15%	80%
Deduct: 3" thick Brick (if available)			M pcs	4%	7%
Add: Ea. Addl. 10' in Fl. Hgt. (or floor)			M pcs	3%	-
Add: Hauling & Unloading if required			M pcs		See Table 4B
Add: Breakage & Cutting Allowance			-	-	3%
Add: Piers & Corbels			M pcs	15%	-
Add: Sills & Soldiers			M pcs	20%	-
Add: Floor Brick (over 2")			M pcs	10%	5%
Add: Weave & Herringbone Pattern			M pcs	20%	7%
Add: Stack bond			M pcs	8%	-
Add: Circular or Radius Work			M pcs	20%	-
Add: Rock Faced			M pcs	10%	-
Add: Two-Face Joint Striking			M pcs	15%	-
Add: Arches (including Formwork)			M pcs	75%	20%
Add: Sawing (Special or Excessive)			M pcs	25%	5%
Add: Winter Work (below 40 degrees)					
	Production Loss		M pcs	10%	-
	Encls - Wall Area - 1 Side & 3 Uses		Sq Ft	.20	.18
	2 Sides & 3 Uses		Sq Ft	.30	.30
	Heaters & Fuel - Wall Area		Sq Ft	.10	.20
	Heat Mortar		Cu Yd	10.00	5.25
Deduct: Residential Work			-	10%	-

Crew - 4 Bricklayers to 3 Tenders (Incl. Mortar Mixers & Operators).
See Section 0905.7 for Thin Veneer Brick and Stone (under 2").
See Section 0206.2 for Paver Brick for Sitework.

DIVISION #4 - MASONRY

0402.0 CONCRETE BLOCK (Bricklayers)

		UNIT	LABOR	MATERIAL	
				StdWt	LtWt

.1 CONVENTIONAL
.11 Foundation or Not Struck - Deduct .30 from Labor Units below.
.12 Backup or Struck One Face - Deduct .15 from Labor Units below.
.13 Partition or Struck Two Faces Used for Labor Units below:

Description	UNIT	LABOR	StdWt	LtWt
12" x 8" x 16" Plain	Ea	2.24	1.55	2.04
Double or Single Corner	Ea	2.27	1.65	2.14
Bull Nose (Sgl or Dbl)	Ea	2.35	1.81	2.29
Bond Beam	Ea	2.30	1.72	2.19
Header	Ea	2.30	1.72	2.19
12" x 8" x 8" Half Length Block	Ea	2.14	1.55	2.04
12" x 4" x 16" Half High Block	Ea	2.14	1.59	2.09
12" x 8" x 8" Lintel	Ea	2.35	1.71	2.19
12" x 16" x 8" Lintel	Ea	2.50	1.76	2.24
12" x 8" x 16" Control Joint	Ea	2.35	1.82	2.34
Half Control Joint	Ea	2.15	1.82	2.29
Fire Rated (4 Hr.)	Ea	2.25	1.82	2.29
L Corner	Ea	2.35	1.65	2.09
Solid or Cap	Ea	2.32	1.87	2.29
8" x 8" x 16" Plain	Ea	2.10	1.16	1.52
Double or Single Corner	Ea	2.15	1.26	1.62
Bullnose (Sgl or Dbl)	Ea	2.17	1.43	1.80
8" x 8" x 8" Half Bullnose	Ea	2.12	1.43	1.80
8" x 8" x 16" Bond Beam	Ea	2.17	1.43	1.68
Header	Ea	2.16	1.43	1.68
8" x 8" x 8" Half Length Block	Ea	2.08	1.16	1.43
8" x 8" x 16" Half High Block	Ea	2.08	1.16	1.60
8" x 8" x 8" Lintel	Ea	2.15	1.36	1.70
8" x 16" x 8" Lintel	Ea	2.35	1.36	1.75
8" x 8" x 16" Control Joint	Ea	2.15	1.46	1.70
8" x 8" x 8" Half Control Joint	Ea	2.05	1.46	1.70
8" x 8" x 16" Fire Rated (2 Hr.)	Ea	2.10	1.30	1.64
(4 Hr.)	Ea	2.10	1.46	1.80
Solid or Cap	Ea	2.15	1.35	1.65
6" x 8" x 16" Plain	Ea	2.02	.94	1.28
Bull Nose (Sgl or Dbl)	Ea	2.04	1.26	1.55
6" x 8" x 8" Half Bullnose	Ea	2.00	1.26	1.55
6" x 8" x 16" Bond Beam or Lintel	Ea	2.06	1.14	1.45
6" x 8" x 8" Half Length Block	Ea	2.00	.98	1.28
6" x 4" x 16" Half High Block	Ea	2.00	1.02	1.35
6" x 8" x 16" Fire Rated (2 Hr.)	Ea	2.02	1.10	1.38
L Corner	Ea	2.20	1.20	1.52
Solid or Cap	Ea	2.05	1.29	1.63
4" x 8" x 16" Plain	Ea	1.95	.76	1.08
Bull Nose	Ea	1.98	1.04	1.30
Bond Beam or Lintel	Ea	2.00	.93	1.21
4" x 8" x 8" Half Length Block	Ea	1.90	.78	1.16
4" x 8" x 16" Half High Block	Ea	1.90	.83	1.12
L Corner	Ea	2.15	.98	1.27
Solid or Cap	Ea	2.02	1.08	1.35

DIVISION #4 - MASONRY

0402.0 CONCRETE BLOCK (Bricklayers)

		UNIT	LABOR	MATERIAL StdWt	LtWt
.1	CONVENTIONAL				
.13	Partition or Struck Two Faces, Cont'd...				
	16" x 8" x 16" Plain	Ea	2.60	1.90	2.55
	16" x 8" x 8" Half Length	Ea	2.45	1.89	2.55
	14" x 8" x 16" Plain	Ea	2.30	1.88	2.50
	14" x 8" x 8" Half Length	Ea	2.16	1.86	2.50
	14" x 8" x 16" Bond Beam	Ea	2.20	2.06	2.65
	10" x 8" x 16" Plain	Ea	2.10	1.45	1.93
	10" x 8" x 8" Half Length	Ea	2.05	1.45	1.93
	10" x 8" x 16" Bond Beam	Ea	2.15	1.60	2.08
	3" x 8" x 16" Plain	Ea	1.93	.78	1.00
	Solid	Ea	2.00	.80	1.30
	Add or Deduct to all Blockwork Above for:				
	Add: Full Size Block	Ea	.08	.08	-
	Add: Breakage and Cutting	Ea	.10	3%	3%
	Add: Jamb and Sash Block	Ea	.15	.10	.10
	Add: Stack Bond Work	Ea	.15	-	-
	Add: Radius or Circular Work	Ea	.90	-	-
	Add: Pilaster, Pier or Pedestal Work	Ea	.42	-	-
	Add: Knock-out Block	Ea	.42	.10	.10
	Deduct: Light Weight Block-Installation	Ea	.10	-	-
	Add: Winter Production Loss (below 40°)	Ea	10%	-	-
	Add: Enclosures - Unit	Ea	.19	.08	.08
	Add: Enclosures - Wall Area 1 Side	SqFt	.22	.16	.16
	Add: Enclosures - Wall Area 2 Sides	SqFt	.32	.26	.26
	Add: Heating & Fuel - Wall Area	SqFt	.10	.18	.20
	Deduct: Residential Work	Ea	10%	-	-

See 0410 for Core Filling for Reinforced Masonry
See 0411 for Filling for Insulated Masonry

		UNIT	LABOR	StdWt	LtWt
.2	ORNAMENTED OR SPECIAL EFFECT				
.21	Shadowall and Hi-Lite Block				
	Add to Block Prices 0402.1	Ea	.14	.12	.13
.22	Scored Block				
	Add to Block Prices 0402.1 - 1 Score	Ea	.19	.10	.10
	- 2 Score	Ea	.25	.17	.17
.23	Break Off Block				
	Add to Block Prices 0402.1 - Broken	Ea	.17	.67	.67
	- Unbroken	Ea	.24	.62	.62
.24	Split Face Block				
	Add to Block Prices 0402.1	Ea	.17	.27	.29
.25	Rock Faced Block				
	Add to Block Prices 0402.1	Ea	.17	.18	.18
.26	Adobe (or Slump) Block				
	Add to Block Prices 0402.1				
	8" x 8" x 16"	Ea	.21	.31	-
	4" x 8" x 16"	Ea	.19	.23	-
.3	SCREEN WALL - 4" x 12" x 12"	Ea	2.06	1.75	-
.4	INTERLOCKING				
.41	Epoxy Laid				
	Add to Block Prices 0402.1	Ea	-	.13	.13
	Deduct from Labor Costs 0402.1	Ea	.41	-	-
.42	Panelized (No Head Joint Mortar)				
	Add to Block Prices 0402.1	Ea	-	.13	.13
	Deduct from Labor Costs 0402.1	Ea	.46	-	-

DIVISION #4 - MASONRY

0402.0 CONCRETE BLOCK, Cont'd...

		UNIT	LABOR	MATERIAL
.5	**SOUND BLOCK**			
	Add to Block Prices 0402.1	Ea	.27	.80
	Add for Fillers	Ea	.16	.27
.6	**BURNISHED BLOCK** - One Face Finish			
	(Truckload Price)			
	Plain 12" x 8" x 16"	Ea	2.55	4.10
	8" x 8" x 16"	Ea	2.35	3.60
	6" x 8" x 16"	Ea	2.24	2.40
	4" x 8" x 16"	Ea	2.14	2.20
	2" x 8" x 16"	Ea	2.10	2.10
	Shapes 12"	Ea	2.93	4.80
	8"	Ea	2.83	3.90
	6"	Ea	2.78	2.70
	4"	Ea	2.68	2.45
	2"	Ea	2.57	2.35
	Add for Bond Beam	Ea	.37	.90
	Add to above for Two Face Finish	Ea	.47	1.70
	Add to above for Scored Face	Ea	.32	.12
	Add for Color	Ea	-	.30
	Add for Bullnose	Ea	.37	.75
.7	**PREFACED UNITS** - Ceramic Glazed			
	(Truckload Price)			
	12" x 8" x 16" Stretcher	Ea	3.20	6.25
	Glazed 2 Face	Ea	3.73	10.00
	8" x 8" x 16" Stretcher	Ea	3.04	5.50
	Glazed 2 Face	Ea	3.34	9.20
	6" x 8" x 16" Stretcher	Ea	3.10	5.25
	Glazed 2 Face	Ea	3.30	9.60
	4" x 8" x 16" Stretcher	Ea	3.20	5.00
	Glazed 2 Face	Ea	3.20	8.80
	2" x 8" x 16" Stretcher	Ea	2.90	5.00
	4" x 16" x 16" Stretcher	Ea	5.40	17.00
	Add for Scored Block	Ea	.32	.40
	Add for Base Caps, Jambs, Headers	Ea	.58	1.75
	Add for Raked Joints and White Cement	Ea	.68	.13
	Add for Less than Truckload Lot	Ea	-	10%

Crew Ratio - 1 Block Layer to 1 Tender
(Tender includes Mixers & Operators)

Deduct 10% from Block Labor for Residential Construction

0403.0 CLAY BACKING AND PARTITION TILE
(Carload Price)

	UNIT	LABOR	MATERIAL
3" x 12" x 12"	Sq Ft	1.70	1.70
4" x 12" x 12"	Sq Ft	1.75	1.75
6" x 12" x 12"	Sq Ft	1.90	2.18
8" x 12" x 12"	Sq Ft	2.00	2.60

DIVISION #4 - MASONRY

		UNIT	LABOR	MATERIAL
0404.0	**CLAY FACING TILE**			
.1	6T SERIES (SELECT QUALITY - CLEAR & FUNCTIONAL)			
	2" x 5-1/3" x 12" Soap Stretcher (Open Back)	Ea	2.27	2.92
	Soap Stretcher (Solid Back)	Ea	2.33	3.02
	4" x 5-1/3" x 12" 1 Face Stretcher	Ea	2.43	4.10
	Select 2 Face Stretcher	Ea	2.58	5.25
	6" x 5-1/3" x 12" Select 1 Face Bond Stretcher	Ea	2.43	4.50
	8" x 5-1/3" x 12" Select 1 Face Bond Stretcher	Ea	2.58	5.60
	Shapes- Group 1 Square and B/N Corners	Ea	2.64	4.55
	2 Sills and 2" Base	Ea	2.71	4.86
	3 Sills and 4" Base	Ea	2.74	6.02
	4 Starters and Octagons	Ea	3.00	11.30
	5 Radials and End Closures	Ea	3.00	18.90
	Add for Less than Carload Lots	Ea	-	10%
	Add for Designer Colors	Ea	-	20%
	Add for Base Only	Ea	-	40%
.2	8W SERIES (SELECT QUALITY - CLEAR & FUNCTIONAL)			
	2" x 8" x 16" Soap Stretcher (Open Back)	Ea	3.07	4.96
	Soap Stretcher (Solid Back)	Ea	3.12	5.35
	4" x 8" x 16" 1 Face Stretcher	Ea	3.13	4.60
	Select 2 Face Stretcher	Ea	3.32	8.65
	6" x 8" x 16" Select 2 Face 6" Bond Stretcher	Ea	3.27	7.35
	8" x 8" x 16" Select 1 Face 8" Bond Stretcher	Ea	3.38	8.95
	Shapes- Group 1 Square and B/N Corners	Ea	3.28	7.75
	2 Sills and 2" Base	Ea	3.80	10.05
	3 Sills and 4" Base	Ea	3.90	12.50
	4 Starters and Octagons	Ea	4.24	15.35
	5 Radials and End Closures	Ea	4.35	37.80
	Add for Less than Carload Lots	Ea	-	40%
	Add for Designer Colors	Ea	-	20%
	Add for Base Only	Ea	25%	-
0405.0	**GLASS UNITS**			
	4" x 8" x 4"	Ea	3.18	5.15
	6" x 6" x 4"	Ea	3.40	5.20
	8" x 8" x 4"	Ea	3.78	5.85
	12" x 12" x 4"	Ea	4.96	17.00
	Add for Solar Reflective Type	Ea	-	40%
	Accessories: Reinforcing & Expansion Joint	Ea	-	1.80
0406.0	**TERRA-COTTA**			
	Unglazed	SqFt	2.22	5.30
	Glazed	SqFt	2.53	6.50
	Colored Glaze	SqFt	2.58	9.40
0407.0	**MISCELLANEOUS UNITS**			
.1	COPING TILE - 9" Double Slant	LnFt	3.50	8.50
	12" Double Slant	LnFt	3.80	11.20
	Bell	LnFt	3.80	9.35
	Corners & Ends = 4 x Above Material Prices			
.2	FLUE LINING - 8" x 8"	LnFt	3.34	4.40
	8" x 12"	LnFt	3.45	8.00
	12" x 12"	LnFt	3.85	9.50
	16" x 16"	LnFt	6.82	14.00
	18" x 18"	LnFt	7.75	15.00
	20" x 20"	LnFt	12.75	29.20
	24" x 24"	LnFt	17.00	37.30

DIVISION #4 - MASONRY

0408.0 NATURAL STONE (M) (Bricklayers)
(See 0906.0 for Stone under 2")

		SQUARE FOOT		CUBIC FOOT	
		Labor	Matl.	Labor	Matl.
.1	CUT STONE				
.11	Limestone				
	Indiana (Standard) and Alabama				
	Face Stone 3"	7.75	16.50	31.00	66.00
	4"	8.70	18.00	26.00	54.00
	Sills and Light Trim	-	-	34.00	65.00
	Copings and Heavy Sections	-	-	30.80	63.00
	Add for Select	2.46	-	12.50	
	Minnesota, Texas, Wisconsin, etc.				
	Face Stone 3"	7.75	22.00	31.00	88.00
	4"	8.70	24.00	26.00	72.00
	Sills and Light Trim	-	-	27.50	87.00
	Copings and Heavy Sections	-	-	26.25	84.50
.12	Marble				
	Face Stone 2"	8.05	27.00	34.00	102.00
	3" and 4"	8.75	29.00	30.80	92.40
	Sills and Light Trim	-	-	32.00	96.00
	Copings and Heavy Sections	-	-	29.00	81.00
.13	Granite				
	Face Stone 2"	8.25	24.00	43.50	135.00
	3"	9.85	26.00	39.40	118.00
	4"	11.45	30.00	34.50	103.50
	Sills and Light Trim	-	-	42.40	101.50
	Copings and Heavy Sections	-	-	39.30	96.00
.14	Slate 1 1/2"	8.75	26.00	52.00	140.00
.2	ASHLAR STONE (40 SqFt/Ton - 4" sawed bed)				
.21	Limestone				
	Indiana - Random	8.15	8.80		
	Coursed (2"-5"-8")	8.48	10.00		
	Minn., Ala., Texas, Wisc., etc.)				
	Split Face-1 Size 7 1/2"	8.70	9.80		
	Coursed (2"-5"-8")	10.10	10.80		
	Sawed/Planed Face-Random	8.32	10.90		
	Coursed (2"-5"-8")	8.05	12.00		
.22	Marble- Sawed Random	9.05	17.50		
	Coursed	9.25	18.00		
.23	Granite- Bushhammered-Random	9.55	20.00		
	Coursed	9.88	21.00		
.3	ROUGH STONE (30 s.f. per Ton)				
.31	Flagstone and Rubble 2"	9.05	8.00		
.32	Field Stone or Boulders	8.50	7.50		
.33	Lt.Wt. Boulders (Igneous) (260 s.f. Ton)				
	2" to 5" Veneer (Sawed Back)	7.15	7.00		
	3" to 10" Boulders	8.50	6.80		
.34	Lt.Wt. Limestone (90 s.f. Ton)				
	2" to 4"	8.15	8.50		

0409.0 PRECAST VENEERS AND SIMULATED MASONRY

		Labor	Matl.
.1	ARCHITECTURAL PRECAST STONE		
	Limestone and Gravel	7.50	17.50
	Marble	8.20	24.60
	Granite and Quartz	8.25	22.80
.2	PRECAST CONCRETE	7.30	13.50
.3	MOSAIC GRANITE PANELS	10.35	28.00

DIVISION #4 - MASONRY

		UNIT	LABOR	MATERIAL
0410.0	**MORTARS**			
	(See Table 4B for Quantities per Unit)			
	<u>Material Cost used for Units Below</u> (Truckload Prices)			
	Portland Cement - Gray	Bag- 94#	9.80	9.20
	White	Bag- 94#	21.10	18.00
	Masonry Cement	Bag- 70#	9.50	7.00
	High Early Cement	Bag- 94#	-	10.00
	Lime, Hydrated	Bag- 50#	5.80	6.10
	Silica Sand	Bag-100#	-	6.70
	Fire Clay	Bag-100#	-	19.70
	Sand	CuYd	32.80	20.20
	Add for Less than Truckload	-	-	20%
	Labor Incl in Unit Costs:w/ Mortar Mixer	CuYd	23.20	-
	Hand Mixed	CuYd	53.00	-
.1	**PORTLAND CEMENT AND LIME MIX**			
	(4.5 Bags Portland and 4.5 Bags Lime)			
	1:1:6 - Gray - Truckload	CuYd	24.40	82.00
	Less than Truckload	CuYd	24.40	96.00
	White	CuYd	24.40	110.00
	Add for Using Silica Sand	CuYd	20.60	33.00
.2	**MASONRY CEMENT MIX (9 Bags Cement)**			
	1:3 - Truckload	CuYd	24.40	67.00
	Less than Truckload	CuYd	24.40	78.00
.3	**EPOXY CEMENT**	CuYd	32.00	163.00
.4	**FIRE RESISTANT CEMENT** (Clay Mix 300# Mpc)	CuYd	32.00	92.00
.5	**ADMIXTURES** (Add to Mortar Prices above)			
	Waterproofing (1# sack cmt & $ .40/lb)	CuYd	16.60	8.25
	Accelerators (1 qt sack cmt & $1.00/lb)	CuYd	16.60	11.50
.6	**COLORING** (50# CuYd or 90# Mpc & $1.00/lb)	CuYd	18.80	45.00
	Standard Brick	M Pcs	23.00	95.00
	Add for Heating Mortar Materials	CuYd	10.60	6.80
	10% allowed in above Prices for Shrinkage			

0411.0 CORE FILLING FOR REINFORCED CONCRETE MASONRY
Based on 3500# Conc.@ $71 CuYd; Mortar @ $70 CuYd; and $23 CuYd to Mix
Deduct 50% from Labor Units if Truck Chuted or Pumped

	Void			Ready Mix		Job Mix	
	CuFt	%	Unit	L	M	L	M
12" x 8" x 16" Plain - 2 Cell	.43	48	Pc	.38	1.16	.72	1.39
	.48	53	SqFt	.40	1.28	.78	1.55
8" x 8" x 16" Plain - 2 Cell	.27	40	Pc	.29	.74	.50	.88
	.30	44	SqFt	.30	.82	.54	.98
6" x 8 " x16" Plain - 2 Cell	.19	40	Pc	.25	.54	.38	.64
	.21	44	SqFt	.26	.57	.41	.68
12" x 8" x 16" Bond Beam	.45	50	Pc	.40	1.23	.80	1.45
	.49	55	SqFt	.42	1.28	.86	1.57
8" x 8" x 16" Bond Beam	.21	32	Pc	.26	.58	.39	.70
	.23	35	SqFt	.27	.62	.41	.75
6" x 8" x 16" Bond Beam	.11	25	Pc	.20	.32	.28	.37
	.13	27	SqFt	.21	.34	.28	.43
12" x 16" x 8" Lintel	.52	58	pc	.47	1.39	.82	1.60
	.57	63	SqFt	.50	1.48	.90	1.75
8" x 16" x 8" Lintel	.27	40	Pc	.29	.74	.47	.86
	.30	44	SqFt	.31	.80	.51	.94
16" x 8" x 18" Pilaster Block	.96	72	Pc	.82	2.56	2.17	3.02
	1.08	81	LnFt	.90	2.84	2.50	3.45
Add for Vertical Rein.			Pc	.14	.15	.15	.16
(1 #4 Rods 16" O.C.)			SqFt	.15	.16	.16	.16

DIVISION #4 - MASONRY

0412.0 CORE & CAVITY FILL FOR INSULATED MASONRY

		R Value-No Insulation	R Value- Insulated	UNIT	LABOR	MATERIAL
.1	LOOSE					
.11	Core Fill					
	12" Concrete Block	1.28	3.90	Pc	.32	.42
				SqFt	.35	.45
	Lt. Wt. Concrete Block	2.27	8.30	Pc	.32	.42
				SqFt	.35	.45
	10" Concrete Block	1.20	2.40	Pc	.28	.36
				SqFt	.30	.39
	Lt. Wt. Concrete Block	2.22	5.70	Pc	.27	.37
				SqFt	.30	.41
	8" Concrete Block	1.11	1.93	Pc	.22	.23
				SqFt	.23	.24
	Lt. Wt. Concrete Block	2.18	5.03	Pc	.22	.23
				SqFt	.23	.24
	6" Concrete Block	.90	1.79	Pc	.20	.16
				SqFt	.21	.17
	Lt. Wt. Concrete Block	1.65	2.99	Pc	.20	.16
				SqFt	.21	.17
			R Value			
	8" Brick		.88	SqFt	.21	.12
	6" Brick		.66	SqFt	.17	.09
	4" Brick		.44	SqFt	.15	.07
	Add to above for Expanded Mica	-	-		10%	200%
	Add to above for Perlite	-	-		-	100%
	10% added for spillage					
.12	Cavity Fill - Per Inch Wall					
	Expanded Styrene		3.8	SqFt	.22	.20
	Fiber Glass		2.2	SqFt	.22	.23
	Rock Wool		2.9	SqFt	.22	.23
	Cellulose		3.7	SqFt	.22	.21
	Add per added inch		-	-	40%	100%
	10% added for spillage					
.2	RIGID					
	Fiber Glass					
	1" - 3# Density		4.35	SqFt	.29	.36
	1.5"		6.52	SqFt	.32	.54
	2"		8.70	SqFt	.36	1.00
	Expanded Styrene - Molded (White)					
	1" - 1# Density		4.33	SqFt	.29	.23
	1.5"		6.52	SqFt	.32	.33
	2"		8.70	SqFt	.36	.57
	Add for Embedded Nailer Type		-	-	-	.10
	Expanded Styrene - Extruded (Blue)					
	1" - 2# Density		5.40	SqFt	.29	.38
	1.5"		8.10	SqFt	.32	.53
	2"		10.80	SqFt	.36	.75
	Expanded Urethane					
	1"		7.14	SqFt	.31	.83
	2"		14.28	SqFt	.36	1.60
	Perlite					
	1"		2.78	SqFt	.28	.43
	2"		5.56	SqFt	.35	.85
	Add for Glued Applications		-	SqFt	.05	.02

DIVISION #4 - MASONRY

0413.0 ACCESSORIES & SPECIALTIES - MATERIAL

	UNIT	LABOR	MATERIAL
.1 ANCHORS			
Dovetail-Brick 12 ga Galvanized x 3 - 1/2"	Ea	.17	.43
Brick 16 ga Galvanized x 3 - 1/2"	Ea	.17	.28
Brick 16 ga Galvanized x 5"	Ea	.18	.36
Stone - Stainless	Ea	.18	1.05
Rigid Wall - 1 - 1/4" x 3/6" x 8" - Galv	Ea	.17	.95
Stic - Kips and Clamp	Ea	.22	.20
Strap Z - 1/8" x 1" x 6" or 8" - Galv	Ea	.36	1.10
1/8" x 1" x 6" - SS	Ea	.36	2.40
3/16" x 1 - 1/4" x 8" - Galv	Ea	.36	1.60
Split Tail - 3/16" x 1 - 1/4" x 6" - Galv	Ea	.36	1.55
1/8" x 1 - 1/2" x 6" - SS	Ea	.36	3.25
Wall Clips - 2" x 3"	Ea	.36	.19
Wall Plugs - Wood Filled	Ea	.36	.21
.2 TIES - WIRE			
Galv Z - 3/16" Wire - 6" & 8"	Ea	.17	.17
Zinc Z - 3/16" Wire - 6" & 8"	Ea	.17	.38
Copperweld Z - 3/16" Wire - 6"	Ea	.17	.44
3/16" Wire - 8"	Ea	.17	.52
Adjustable Rect. - 3 /16" Wire 3 - 1/4"	Ea	.17	.30
3/16" Wire 4 - 3/4"	Ea	.17	.34
Corrugated-Strap - 26 ga - 7/8" x 7"	Ea	.17	.04
22 ga - 7/8" x 7"	Ea	.17	.05
16 ga - 7/8" x 7"	Ea	.17	.08
Hardware Cloth - 2" x 5" x 1/4" - 23 ga	Ea	.17	.14
23 ga x 1/4" x 30"	SqFt	.24	.60
.3 REINFORCEMENT			
#9 Welded Wire - Ladder Type - 4"	LnFt	.15	.19
6"	LnFt	.15	.20
8"	LnFt	.16	.22
10"	LnFt	.16	.23
12"	LnFt	.18	.26
16"	LnFt	.20	.36
Add for 3/16" Wire H.D.	LnFt	-	30%
Add for Truss Type	LnFt	-	20%
Add for Rect. Ties - Composite & Cavity	LnFt	.03	.06
Add for Rect. Adj. Ties - Comp & Cavity	LnFt	.03	.09
Reinforcement Rods - Horizontal Placement			
3/8" or #3 .376 lb	LnFt	.16	.15
1/2" or #4 .668 lb	LnFt	.21	.20
5/8" or #5 1.043 lb	LnFt	.32	.30
Add for Vertical Placement	LnFt	25%	-
.4 FIREPLACE AND CHIMNEY ACCESSORIES			
Ash Dumps and Thimbles - 6" x 8"	Ea	27.00	12.00
Clean Cut Doors, Cast Iron - 8" x 8"	Ea	25.00	20.00
12" x 12"	Ea	30.00	42.00
Dome Dampers, Cast Iron - 24" x 16"	Ea	22.00	48.00
30" x 16"	Ea	26.00	63.00
36" x 20"	Ea	38.00	90.00
42" x 20"	Ea	50.00	95.00
48" x 24"	Ea	70.00	130.00

DIVISION #4 - MASONRY

		UNIT	LABOR	MATERIAL	
				1/2"	1"
0413.0	**ACCESSORIES & SPECIALTIES, Cont'd...**				
.5	CONTROL AND EXPANSION JOINT MATERIAL				
	Asphalt - Fibre	SqFt	.53	1.00	1.85
	Polyethylene Foam	SqFt	.53	1.10	1.95
	Sponge Rubber	SqFt	.54	6.25	7.70
	Paper Fibre	SqFt	.50	.80	1.50
	Also in Thk. of 1/8", 1/4", 3/8" & 3/4"				
				2 oz	3 oz
.6	FLASHING, THRU WALL				
	Copper Between Polyethylene	SqFt	.70	1.16	1.45
	Rein Paper - 2 oz/1.05	SqFt	.65	1.31	2.00
	Asphalted Fabric - 3 oz	SqFt	.65	1.73	2.40
	Poly Vinyl Chloride - 20 mil	SqFt	.55	-	.22
	30 mil	SqFt	.60	-	.38
	Add for Adhesives			.05	.05
0414.0	**CLEANING & POINTING** (See 0414 for Restoration)				
.1	BRICK & STONE- Acid & Other Chemicals	SqFt	.40	-	.03
	Soap & Water	SqFt	.35	-	-
.2	BLOCK - 1 Face	SqFt	.20	-	-
.3	FACING TILE AND BLOCK				
	Point - White Cement and Sand	SqFt	.70	-	.08
	Silica Sand	SqFt	.85	-	.12
	Add for Scaffold (See 0417)				
0415.0	**MASONRY RESTORATION (L&M) (Bricklayers)**				
					Cost
.1	RAKING, FILLING AND TUCKPOINTING				
	Limited - Standard Brick & 3/8" Joint	SqFt			2.80
	Heavy	SqFt			3.50
	All	SqFt			6.80
	Deduct for Stone	SqFt			20%
	Deduct for 4" x 12" Brick	SqFt			30%
	Add for Larger & Smaller Joints	SqFt			25%
.2	SAND BLAST				
.22	Brick -Wet	SqFt			1.55
	Dry	SqFt			1.45
.23	Block -Wet	SqFt			1.45
	Dry	SqFt			1.33
.3	STEAM CLEAN AND CHEMICAL CLEANING	SqFt			1.30
.4	HIGH PRESSURE WATER CLEANING (Hydro)	SqFt			.90
.5	BRICK REPLACEMENT -Standard	Each			24.00
	4" x 12"	Each			19.00
	Add for Scaffolding	SqFt			.66
	Add for Protection	SqFt			.28
.6	WATER PROOFING - CLEAR	SqFt			.80
	See 0702.3 for Waterproofing Masonry				
	See 0711.0 for Caulking				
0416.0	**INSTALLATION OF STEEL & MISC. EMBEDDED ITEMS**				
	See Division 0502				
0417.0	**EQUIPMENT AND SCAFFOLD**				
.1	EQUIPMENT AVERAGE OF LABOR				
	See 1-7A for Rental Rates		LABOR	MATERIAL	
.2	SCAFFOLD - INCLUDING PLANK				
	Tubular Frame- to 40' - Exterior	SqFt	.36	.20	
	to 16' - Interior	SqFt	.33	.20	
	Swing Stage (Incl. Outriggers & Anchorage)	SqFt	.32	.22	

DIVISION #4 - MASONRY - QUICK ESTIMATING

The costs below are average but can vary greatly with Quantity, Weather, Area Practices and many other factors. They are priced as total contractor's price with an overhead and fee of 10% included. Units include mortar, ties, cleaning, and reinforcing. Not included are scaffold, hoisting, lintels, accessories, heat, and enclosures. Also added are 27% Insurance on Labor; 10% for General Conditions, Equipment and Tools; and 5% on Material for Sales Taxes. Crew Size is based on 4 Bricklayers to 3 Tenders for Brick and Stone Work, and 4 Bricklayers to 4 Tenders for Block and Tile. See Table 3 for Production Standards. Deduct 5% for Residential Work.

Example:	.1	Face Brick 8" x 2 2/3" x 4"	UNIT	LABOR	MATERIAL	TOTAL
	.11	Running Bond	Each	.70	.50	1.20
		Mortar and Clean Brick	Each	.10	.10	.20
				.80	.60	1.40
		Add: 27% Labor Burden, 5% Sales Tax		.20	.03	.23
				1.00	.63	1.63
		Add: 10% General Conditions & Equipment				.16
						1.79
		Add: 10% Overhead and Fee				.18
		Total Brick Cost	Each Unit			1.97
			or SqFt			13.30

				COST	
0401.0	**BRICK MASONRY**				
.1	FACE BRICK			SqFt	Unit
.11	Conventional (Modular Size)				
	Running Bond - 8" x 2 2/3" x 4"			13.30	1.97
	Common Bond - 6 Course Header			15.02	1.98
	Stack Bond			13.77	2.04
	Dutch & English Bond - Every Other Course Header			21.28	2.10
	Every Course Header			23.68	2.10
	Flemish Bond - Every Other Course Header			15.85	2.10
	Every Course Header			18.90	2.10
	Add: If Scaffold Needed			.72	.11
	Add: For each 10' of Floor Hgt. or Floor (3%)			.34	.05
	Add: Piers and Corbels (15% to Labor)			1.14	.17
	Add: Sills and Soldiers (20% to Labor)			1.35	.20
	Add: Floor Brick (10% to Labor)			.72	.11
	Add: Weave & Herringbone Patterns (20% to Labor)			1.48	.22
	Add: Stack Bond (8% to Labor)			.60	.09
	Add: Circular or Radius Work 20% to Labor)			1.48	.22
	Add: Rock Faced & Slurried Face (10% to Labor)			.72	.11
	Add: Arches (75% to Labor)			4.72	.70
	Add: For Winter Work (below 40°)				
	Production Loss (10% to Labor)			.72	.11
	Enclosures - Wall Area Conventional			.72	.11
	Heat and Fuel - Wall Area Conventional			.27	.04
.12	Econo - 8" x 4" x 3"			11.16	2.48
.13	Panel - 8" x 8" x 4"			9.68	4.30
.14	Norman - 12" x 2 2/3" x 4"			10.44	2.32
.15	King Size - 10" x 2 5/8" x 4"			9.40	2.00
.16	Norwegian - 12" x 3 1/5" x 4"			9.38	2.50
.17	Saxon-Utility - 12" x 4" x 3"			9.00	3.00
	12" x 4" x 4"			9.15	3.05
	12" x 4" x 6"			11.70	3.90
	12" x 4" x 8"			14.64	4.88
.18	Adobe - 12" x 3" x 4"			10.20	2.55
.2	COATED BRICK (Ceramic)	10.56			2.75
.3	COMMON BRICK (Clay, Concrete and Sand Lime)			9.45	1.40
.4	FIRE BRICK - Light Duty			17.30	2.60
	Heavy Duty			20.48	3.05
	Deduct for Residential Work - All Above			-	5%

DIVISION #4 - MASONRY - QUICK ESTIMATING

		UNIT	LABOR	MATERIAL	TOTAL
Example:	12" x 8" x 16" Concrete Block - Partition	Each	2.24	1.55	3.79
	Mortar (Labor in Block Placing Above)	Each	-	.20	.20
	Reinforcing Wire 16" O.C. Horizontal	Each	.09	.08	.17
	Clean Block - 2 Faces	Each	.30	-	.30
			2.63	1.83	4.46
Add:	27% Taxes and Insurance on Labor				.71
Add:	5% Sales Tax on Material				.09
					5.26
Add:	10% General Cond., Equip. and Tools				.52
					5.78
Add:	10% Overhead and Profit				.58
Total Block Cost		Each			6.36
		or SqFt			7.15

Not Included - Scaffold, Hoisting Equipment, Lintels, Embedded Accessories, Insulation, Heat and Enclosures. (See "Adds" Below.)

0402.0 CONCRETE BLOCK

	Std Wt Cost		Lt Wt Cost	
.1 CONVENTIONAL (Struck 2 Sides - Partitions)	SqFt	Unit	SqFt	Unit
12" x 8" x 16" Plain	7.15	6.36	7.48	6.65
Bond Beam (w/ Fill-Reinforcing)	9.75	8.68	10.26	9.12
12" x 8" x 8" Half Block	6.92	6.15	7.24	6.44
Double End - Header	7.25	6.45	8.05	7.17
8" x 8" x 16" Plain	6.13	5.45	6.43	5.72
Bond Beam (w/ Fill-Reinforcing)	7.40	6.58	7.62	6.78
8" x 8" x 8" Half Block	6.00	5.33	6.00	5.35
Double End - Header	6.57	5.84	7.63	6.79
6" x 8" x 16" Plain	5.41	4.81	6.18	5.50
Bond Beam (w/ Fill-Reinforcing)	6.46	5.75	6.34	5.64
Half Block	5.40	4.80	6.15	5.47
4" x 8" x 16" Plain	5.20	4.62	5.19	4.62
Half Block	5.07	4.51	4.96	4.41
16" x 8" x 16" Plain	8.26	7.35	9.07	8.07
Half Block	7.87	7.00	8.93	7.74
14" x 8" x 16" Plain	7.63	6.79	8.34	7.42
Half Block	7.15	6.36	7.98	7.10
10" x 8" x 16" Plain	7.44	5.62	7.21	6.41
Bond Beam (w/ Fill-Reinforcing)	7.15	6.36	8.45	7.52
Half Block	6.30	5.60	6.86	6.10
Deduct: Block Struck or Cleaned One Side	.24	.21	.24	.21
Deduct: Block Not Struck or Cleaned Two Sides	.45	.40	.45	.40
Deduct: Clean One Side Only	.17	.15	.17	.15
Deduct: Lt. Wt. Block (Labor Only)	.15	.14	.15	.14
Add: Jamb and Sash Block	.40	.36	.40	.36
Add: Bullnose Block	.60	.53	.60	.53
Add: Pilaster, Pier, Pedestal Work	.66	.59	.66	.59
Add: Stack Bond Work	.26	.23	.26	.23
Add: Radius or Circular Work	1.18	1.07	1.18	1.07
Add: If Scaffold Needed	.68	.60	.68	.60
Add: Winter Production Cost (10% Labor)	.38	.34	.38	.34
Add: Winter Enclosing - Wall Area	.42	.37	.42	.37
Add: Winter Heating - Wall Area	.33	.29	.33	.29
Add: Core & Beam Filling - See 0411 (Page 4A-18)				
Add: Insulation - See 0412 (Page 4A-18)				

Deduct for Residential Work - 5%

DIVISION #4 - MASONRY - QUICK ESTIMATING

		Std Wt Cost		Lt Wt Cost	
0402.0	**CONCRETE BLOCK, Cont'd...**	SqFt	Unit	SqFt	Unit
.2	ORNAMENTAL OR SPECIAL EFFECT - Add to Units Above in 0402.1				
	Shadow Wall	.35	.31	.37	.33
	Scored Block	.41	.36	.40	.36
	Breakoff	1.27	1.13	1.26	1.12
	Split Face	.52	.46	.67	.60
	Rock Faced	.54	.45	.54	.48
	Adobe (Slump)	.73	.65	-	-
.3	SCREEN WALL - 4" x 12" x 12"	4.90	4.90	-	-
.4	INTERLOCKING - <u>Deduct</u> from Units 0402.1				
	Epoxy Laid	.46	.41	.42	.41
	Panelized	.66	.59	.62	.59
.5	SOUND BLOCK - <u>Add</u> to Block Prices 0402.1			1.56	1.40
.6	BURNISHED - 12" x 8" x 16"			10.80	9.60
	8" x 8" x 16"			6.20	7.30
	6" x 8" x 16"			7.70	6.85
	4" x 8" x 16"			6.95	6.20
	2" x 8" x 16"			6.65	5.90
	Add for Shapes			1.22	1.08
	Add for 2 Faced Finish			3.03	2.70
	Add for Scored Finish - 1 Face			.67	.60
.7	PREFACED UNITS - 12" x 8" x 16" Stretcher			15.75	14.00
	(Ceramic Glazed) 12" x 8" x 16" Glazed 2 Face			22.00	19.60
	8" x 8" x 16" Stretcher			14.45	12.85
	Glazed 2 Face			20.10	17.70
	6" x 8" x 16" Stretcher			13.40	11.90
	Glazed 2 Face			18.45	16.40
	4" x 8" x 16" Stretcher			12.65	11.25
	Glazed 2 Face			19.00	16.90
	2" x 8" x 16" Stretcher			12.15	10.80
	4" x 16" x 16" Stretcher			36.00	32.10
	Add for Base, Caps, Jambs, Headers, Lintels			3.60	3.20
	Add for Scored Block			1.07	.95

		COST	
		SqFt	Unit
0403.0	**CLAY BACKING AND PARTITION TILE**		
	3" x 12" x 12"	4.75	4.75
	4" x 12" x 12"	4.95	4.95
	6" x 12" x 12"	5.90	5.90
	8" x 12" x 12"	6.30	6.30
0404.0	**CLAY FACING TILE (GLAZED)**		
.1	6T or 5" x 12" - SERIES		
	2" x 5 1/3" x 12" Soap Stretcher (Solid Back)	15.40	6.85
	4" x 5 1/2" x 12" 1 Face Stretcher	17.80	7.90
	2 Face Stretcher	21.60	9.60
	6" x 5 1/3" x 12" 1 Face Stretcher	21.80	9.70
	8" x 5 1/3" x 12" 1 Face Stretcher	27.20	12.10
.2	8W or 8" x 16" - SERIES		
	2" x 8" x 16" Soap Stretcher (Solid Back)	13.15	11.70
	4" x 8" x 16" 1 Face Stretcher	13.90	12.35
	2 Face Stretcher	16.65	14.80
	6" x 8" x 16" 1 Face Stretcher	16.40	14.60
	8" x 8" x 16" 1 Face Stretcher	18.05	16.05
	Add for Shapes (Average)	50%	50%
	Add for Designer Colors	-	20%
	Add for Less than Truckload Lots	-	10%
	Add for Base Only	-	25%

4A-16

DIVISION #4 - MASONRY - QUICK ESTIMATING

			COST	
			SqFt	Unit
0405.0	**GLASS UNITS**			
	4" x 8" x 4"		40.40	10.10
	6" x 6" x 4"		42.00	10.50
	8" x 8" x 4"		25.30	11.25
	12" x 12" x 4"		24.50	24.50
0406.0	**TERRA-COTTA**			
	Unglazed		9.80	9.80
	Glazed		13.65	13.65
	Colored Glazed		16.00	16.00
0407.0	**FLUE LINING**			Ln Ft
	8" x 12"			14.70
	12" x 12"			17.50
	16" x 16"			28.00
	18" x 18"			31.00
	20" x 20"			56.00
	24" x 24"			78.00
0408.0	**NATURAL STONE**			
.1	CUT STONE		Sq Ft	Cu Ft
	Limestone- Indiana and Alabama - 3"		33.10	127.60
	4"		35.70	107.00
	Minnesota, Wisc, Texas, etc.- 3"		37.80	151.00
	4"		42.70	128.00
.12	Marble - 2"		45.50	280.00
	3"		47.20	192.00
.13	Granite - 2"		42.20	253.00
	3"		46.00	184.00
.14	Slate - 1 1/2"		42.20	253.00
.2	ASHLAR- 4" Sawed Bed			
.21	Limestone- Indiana - Random		23.80	71.40
	Coursed - 2" - 5" - 8"		25.00	75.00
	Minnesota, Alabama, Wisconsin, etc.			
	Split Face - Random		25.60	76.80
	Coursed		28.30	84.90
	Sawed or Planed Face - Random		27.00	81.00
	Coursed		28.20	84.60
.22	Marble - Sawed - Random		35.70	107.00
	Coursed		36.80	110.40
.23	Granite - Bushhammered - Random		37.80	113.40
	Coursed		40.00	120.00
.24	Quartzite		27.00	81.00
.3	ROUGH STONE			
.31	Rubble and Flagstone		25.00	75.00
.32	Field Stone or Boulders		23.80	71.40
.33	Light Weight Boulders (Igneous)			
	2" to 4" Veneer - Sawed Back		20.60	-
	3" to 10" Boulders		21.70	-
.34	Light Weight Limestone		22.70	-
0409.0	**PRECAST VENEERS AND SIMULATED MASONRY**		Unit	Cost
.1	ARCHITECTURAL PRECAST STONE- Limestone		SqFt	32.00
	Marble		SqFt	38.20
.2	PRECAST CONCRETE		SqFt	25.00
.3	MOSAIC GRANITE PANELS		SqFt	36.30
	See 0904 for Veneer Stones (under 2")			
0410.0	**MORTAR**			
	Portland Cement and Lime Mortar - Labor in Unit Costs		CuYd	114.00
	Masonry Cement Mortar - Labor in Unit Costs		CuYd	103.00
	Mortar for Standard Brick - Material Only		SqFt	.40
	Add for less than Truckload		CuYd	13.00

4A-17

DIVISION #4 - MASONRY - QUICK ESTIMATING

		COST	
		SqFt	Unit
0411.0	**CORE FILLING FOR REINFORCED CONCRETE BLOCK**		
	Add to Block Prices in 0402.0 (Job Mixed)		
	12" x 8" x 16" Plain (includes #4 Rod 16" O.C. Vertical)	2.85	2.54
	Bond Beam (includes 2 #4 Rods)	2.90	2.63
	8" x 8" x 6" Plain (includes #4 Rod 16" O.C. Vertical)	1.97	1.77
	Bond Beam (includes 2 #4 Rods)	1.42	1.40
	6" x 8" x 16" Plain	1.25	1.23
	Bond Beam (includes 1 #4 Rod)	.92	.94
	10" x 8" x 16" Bond Beam (includes 2 #4 Rods)	2.50	2.40
	14" x 8" x 16" Bond Beam (includes 2 #4 Rods)	3.20	3.05
0412.0	**CORE AND CAVITY FILL FOR INSULATED MASONRY**		
	See Page 4-11 for R Values		
.1	LOOSE		
	Core Fill - Expanded Styrene		
	12" x 8" x 16" Concrete Block	.99	.97
	10" x 8" x 16" Concrete Block	.68	.70
	8" x 8" x 16" Concrete Block	.55	.60
	6" x 8" x 16" Concrete Block	.43	.46
	8" Brick - Jumbo - Thru the Wall	.39	.43
	6" Brick - Jumbo - Thru the Wall	.29	.33
	4" Brick - Jumbo	.22	.26
	Cavity Fill - per Inch		
	Expanded Styrene	.50	-
	Mica	.83	-
	Fiber Glass	.52	-
	Rock Wool	.52	-
	Cellulose	.49	-
.2	RIGID - Fiber Glass - 1" 3# Density	.77	-
	2"	1.23	-
	Expanded Styrene - Molded - 1" 1# Density	.56	-
	2"	.81	-
	Extruded - 1" 2# Density	.82	-
	2"	1.30	-
	Expanded Urethane - 1"	1.08	-
	2"	1.83	-
	Perlite - 1"	.81	-
	2"	1.24	-
	Add for Embedded Water Type	.15	-
	Add for Glued Applications	.06	-
0413.0	**ACCESSORIES AND SPECIALTIES**		
	See Page 4-12		
0414.0	**CLEANING AND POINTING**		
	Brick and Stone - Acid or Chemicals	.66	-
	Soap and Water	.60	-
	Block & Facing Tile - 1 Face (Incl. Hollow Metal Frames)	.33	-
	Point with White Cement	1.35	-
0415.0	**MASONRY RESTORATION** - See 0415.0		
0416.0	**INSTALLATION OF STEEL** - See 0502.0		
0417.0	**SCAFFOLD**		
.1	EQUIPMENT, TOOLS AND BLADES		
	Percentage of Labor as an average	-	7%
	See 1 - 7A for Rental Rates & New Costs		
.2	SCAFFOLD - Tubular Frame - to 40 feet - Exterior	.75	-
	Tubular Frame - to 16 feet - Interior	.70	-
	Swing Stage - 40 feet and up	.70	-

DIVISION #4 - MASONRY - QUICK ESTIMATING

TABLE B

		Modular Manufactured Size - Inches (l × h × d)	Mortar CuFt M	Unit Per SqFt	Units Placed ManDay	Haul & Unload M PCs	Unit Wt.
0401.0	**BRICK MASONRY**						
.1	Face Brick						
.11	Conventional						
	Running Bond	7-5/8 x 2-1/4 x 3-5/8	13	6.75	600	16.00	4.5
	Common Bond						
	Header Every 6th Course			ƒ1/6	570	-	-
	Header Every 7th Course			ƒ1/7	570	-	-
	Dutch and English Bond						
	Header Every Other Course			ƒ1/2	510	-	-
	Header Every Course			ƒ2/3	510	-	-
	Flemish Bond						
	Header Every Other Course			ƒ1/8	500	-	-
	Header Every Course			ƒ1/3	500	-	-
.12	Econo	7-5/8 x 3-5/8 x 3-5/8		4.50	430	20.00	6.0
.13	Panel	7-5/8 x 7-5/8 x 3-5/8		2.25	280	40.00	13.5
		7-5/8 x 15-5/8 x 3-5/8		1.25	180	80.00	27.0
.14	Norman	11-5/8 x 2-1/4 x 3-5/8	16	4.50	450	20.00	6.0
.15	King	9-5/8 x 3-1/2 x 3-5/8	16	4.70	440	22.00	4.5
.16	Norwegian	11-5/8 x 2-13/16 x 3-5/8	16	3.75	370	22.00	8.0
.17	Saxon 4"	11-5/8 x 3-5/8 x 3-5/8	19	3.00	320	26.00	9.0
	6"	11-5/8 x 3-5/8 x 5-5/8	28	3.00	260	35.00	13.5
	8"	11-5/8 x 3-5/8 x 7-5/8	38	3.00	200	50.00	18.0

Deduct from above dimensions for 1/2" Joint 1/8" x 1/8" x 1/8".
Add to above dimensions for full size 3/8" x 0" x 2/8".

.2	Coated Brick (Glazed)		12	6.75	510	17.00	4.5
.3	Common Brick						
	Clay	7-14 x 2-1/16 x 3-1/2	13	6.75	660	15.00	4.5
	Concrete		13	6.75	660	15.00	4.5
.4	Fire Brick	8-0 x 2-1/4 x 3-1/2	13	6.65	660	15.00	7.0
0402.0	**STANDARD AND LIGHT-WEIGHT CONCRETE BLOCK**						
	12" Plain	11-5/8 x 7-5/8 x 15-5/8	65	1.125	190	110.00	5.0
	1/2 Height	11-5/8 x 3-5/8 x 15-5/8		2.25		Lt Wt	38.0
	1/2 Length	11-5/8 x 7-5/8 x 7-5/8		2.25			
	8" Plain	7-5/8 x 7-5/8 x 15-5/8	55	1.125	200	80.00	36.0
	1/2 Height	7-5/8 x 3-5/8 x 15-5/8		2.25		Lt Wt	27.0
	1/2 Length	7-5/8 x 7-5/8 x 7-5/8		2.25			
	6" Plain	5-5/8 x 7-5/8 x 15-5/8	45	1.125	210	60.00	28.0
	1/2 Height	5-5/8 x 3-5/8 x 15-5/8		2.25		Lt Wt	21.0
	1/2 Length	5-5/8 x 7-5/8 x 7-5/8		2.25			
	4" Plain	3-5/8 x 7-5/8 x 15-5/8	35	1.125	210	50.00	24.0
	1/2 Height	3-5/8 x 3-5/8 x 15-5/8		2.25		Lt Wt	18.0
	1/2 Length	3-5/8 x 7-5/8 x 7-5/8		2.25			
	16" Plain	15-5/8 x 7-5/8 x 15-5/8	70	2.25	170	150.00	61.0
	14" Plain	13-5/8 x 7-5/8 x 15-5/8	65	2.25	180	90.00	56.0
	10" Plain	9-5/8 x 7-5/8 x 15-5/8	60	2.25	195	85.00	43.0

Average 2 cubic yards mortar per M Block.

4B-19

DIVISION #4 - MASONRY - QUICK ESTIMATING

TABLE B

		Modular Manufactured Size - Inches			Mortar CuFt M	Unit Per SqFt	Units Placed ManDay	Haul & Unload M PCs	Unit Wt.
		l	h	d					
0403.0	**CLAY BACKING & PARTITION TILE**								
	3"	3 x	11-1/2 x	11-1/2	18.7	1.00	260	25.0	13.0
	4"	4 x	11-1/2 x	11-1/2	22.2	1.00	250	29.0	15.0
	6"	6 x	11-1/2 x	11-1/2	16.0	1.00	230	38.0	20.0
0404.0	**CLAY FACING TILE (GLAZED)**								
	2" 6T	1-3/4 x	5-1/16 x	7-3/4	13	2.25	180	20.00	6.5
	4" 6T	3-3/4 x	5-1/16 x	11-3/4		2.25		30.00	10.5
	6" 6T	5-3/4 x	5-1/16 x	11-3/4		2.25		45.00	15.2
	8" 6T	7-3/4 x	5-1/16 x	11-3/4		2.25		55.00	19.0
	2" 8W	1-3/4 x	7-3/4 x	15-3/4	22	1.13	125	40.00	14.0
	4" 8W	3-3/4 x	7-3/4 x	15-3/4		1.13		60.00	21.0
	6" 8W	5-3/4 x	7-3/4 x	15-3/4		1.13		85.00	31.0
0405.0	**GLASS UNITS**								
	6" x 6"	5-3/4 x	5-3/4 x	3-7/8	12.5	4.00	140	18.00	4.0
	6" x 8"	7-3/4 x	7-3/4 x	3-7/8	16.0	2.25	125	26.00	7.0
	12" x 12"	11-3/4 x	11-3/4 x	3-7/8	23.3	1.00	90	33.00	16.0
0406.0	**TERRA-COTTA**								
	3"	3" x	12" x	12"	25	1.00		36.00	10.0
	4"	4" x	12" x	12"	30	1.00		40.00	13.0
	6"	6" x	12" x	12"	40	1.00		47.00	18.0
	8"	8" x	12" x	12"	60				
0407.0	**MISCELLANEOUS UNITS**								
	Coping Tile	9"			50	-	160	35.00	-
		12"			60	-	140	40.00	-
	Flue Lining	8" x	8"		50	-	160	35.00	-
		12" x	12"		55	-	130	45.00	-
		18" x	18"		65	-	60	55.00	-
		21" x	21"		75	-	40	70.00	-
		24" x	24"		90	-	30	90.00	-
0408.0	**NATURAL STONE**								
	Cut Stone				SqFt				CuFt
	Lime Stone				10.0	-	18	25.00	150.0
	Marble				10.0	-	13	25.00	160.0
	Granite				10.0	-	11	25.00	165.0
	Ashlar Stone				15.0	-	26	25.00	150.0
	Rough Stone - Flagstone				15.0	-	22	25.00	140.0
	Field Stone				15.0	-	22	25.00	160.0
	Light Weight Stone				15.0	-	24	25.00	220.0
0409.0	**PRECAST VENEERS**								
	Architectural Precast Stone				10.0	-	18	25.00	150.0
	Terra-Cotta				10.0	-	17	25.00	130.0
	Precast Concrete				10.0	-	24	25.00	150.0

DIVISION #5 - METALS

Wage Rates (Including Fringes) & Location Modifiers
July 2002-2003

	Metropolitan Area		Iron Workers Wage Rate	Wage Rate Location Modifier
1.	Akron	*	43.72	129
2.	Albany-Schenectady-Troy		34.15	102
3.	Atlanta		26.74	80
4.	Austin	**	23.65	71
5.	Baltimore		35.53	95
6.	Birmingham	**	24.54	75
7.	Boston		41.80	127
8.	Buffalo-Niagara Falls		37.92	113
9.	Charlotte	**	21.58	64
10.	Chicago-Gary		46.83	137
11.	Cincinnati		31.97	98
12.	Cleveland		39.34	115
13.	Columbus	*	34.10	102
14.	Dallas-Fort Worth	**	21.84	67
15.	Dayton	*	33.26	100
16.	Denver-Boulder		25.30	80
17.	Detroit		43.68	128
18.	Flint	*	34.65	105
19.	Grand Rapids		34.86	105
20.	Greensboro-West Salem	**	21.58	64
21.	Hartford-New Britain	*	35.25	107
22.	Houston	**	22.83	76
23.	Indianapolis		34.82	106
24.	Jacksonville	**	24.22	73
25.	Kansas City		33.37	101
26.	Los Angeles-Long Beach		39.96	124
27.	Louisville	**	28.77	88
28.	Memphis	**	25.57	80
29.	Miami	**	25.19	76
30.	Milwaukee		37.75	113
31.	Minneapolis-St. Paul		38.76	111
32.	Nashville	**	24.17	75
33.	New Orleans	**	23.47	69
34.	New York		65.90	202
35.	Norfolk-Portsmouth	**	24.81	77
36.	Oklahoma City	**	26.38	82
37.	Omaha-Council Bluffs		29.70	81
38.	Orlando	**	28.26	86
39.	Philadelphia		46.03	143
40.	Phoenix	**	29.78	100
41.	Pittsburgh		39.05	118
42.	Portland		35.22	112
43.	Providence-Pawtucket	*	37.73	115
44.	Richmond	**	24.27	77
45.	Rochester		34.37	109
46.	Sacramento		41.03	127
47.	St. Louis		39.98	107
48.	Salt Lake City-Ogden	**	25.67	77
49.	San Antonio	**	21.20	72
50.	San Diego		42.48	130
51.	San Francisco-Oakland-San Jose		42.27	128
52.	Seattle-Everett		65.15	106
53.	Springfield-Holyoke-Chicopee	*	35.63	114
54.	Syracuse	*	34.62	108
55.	Tampa-St. Petersburg	**	39.77	87
56.	Toledo		34.47	110
57.	Tulsa	**	27.20	82
58.	Tucson	**	32.40	100
59.	Washington D.C.		30.84	90
60.	Youngstown-Warren		33.62	104
	AVERAGE		**33.63**	

Impact Ratio : Labor 30%, Material 70%
* Contract Not Settled - Wage Interpolated
** Non Signatory or Open Shop Rate

DIVISION #5 - METALS

		PAGE
0501.0	**STRUCTURAL METAL FRAMING (05100)**	5-3
.1	BEAMS and COLUMNS	5-3
.2	CHANNELS	5-3
.3	ANGLES	5-3
.4	STRUCTURAL T's	5-3
.5	STRUCTURAL TUBING	5-3
.6	PLATES and BARS	5-3
.7	BOX GIRDERS	5-3
.8	TRUSSES	5-3
0502.0	**MISCELLANEOUS METALS (05500)**	5-3
.1	BOLTS	5-3
.2	CASTINGS (05540)	5-3
.3	CORNER GUARDS	5-3
.4	FIRE ESCAPES	5-3
.5	FRAMES, CHANNEL	5-3
.6	GRATINGS and TRENCH COVERS (05532)	5-3
.7	GRILLES	5-3
.8	HANDRAILS and RAILINGS (05720)	5-3
.9	LADDERS (05515)	5-3
.10	LINTELS, LOOSE	5-3
.11	PIPE GUARDS	5-4
.12	PLATE, CHECKERED	5-4
.13	STAIRS - METAL PAN (05712)	5-4
.14	STAIRS - CIRCULAR (05715)	5-4
.15	WIRE GUARDS	5-4
0503.0	**ORNAMENTAL METALS - NON FERROUS (05760)**	5-4
0504.0	**OPEN WEB JOISTS (05160)**	5-4
.1	STANDARD BAR JOISTS	5-4
.2	LONG SPAN JOISTS	5-4
0505.0	**METAL DECKINGS (05300)**	5-4
.1	SHORT SPAN	5-4
.2	LONG SPAN	5-4
.3	CELLULAR - ELECTRICAL and AIR FLOW	5-5
.4	CORRUGATED	5-5
.5	BOX RIB	5-5
0506.0	**METAL SIDINGS AND ROOFINGS (07411)**	5-5
.1	INDUSTRIAL	5-5
.11	Steel	5-5
.12	Aluminum	5-5
.2	ARCHITECTURAL	5-5
.21	Steel	5-5
.22	Aluminum	5-5
.23	Stainless Steel	5-5
.24	Protected Steel	5-5
	See Section 1310 for Metal Buildings	
0507.0	**LIGHT GAGE COLD-FORMED METAL FRAMING (05400)**	5-5
0508.0	**EXPANSION CONTROL (05800)**	5-5
	See Section 0303 for Reinforcing Steel	

DIVISION #5 - METALS

0501.0 STRUCTURAL METAL (L+M) (Ironworkers)

		UNIT	LABOR	MATERIAL
.1	BEAMS AND COLUMNS			
.11	Wide Flange Shapes, H-Bearing Piles		(Subcontracted)	
	Light Weight Flange Columns & Misc. Columns			
	Light Beams - American Standard			
	Small Job (5 to 30 ton)	Ton	420.00	1,250.00
	Medium Job (30 to 100 ton)	Ton	410.00	1,230.00
	Large Job (100 ton and up)	Ton	380.00	1,180.00
.12	Junior Beams	Ton	395.00	1,280.00
.2	CHANNELS - American Std. - Junior -3" to 15"	Ton	540.00	1,150.00
	10" to 12"	Ton	520.00	1,280.00
.3	ANGLES (Equal and Unequal)	Ton	590.00	1,400.00
.4	STRUCTURAL T's (from wide Flange Shapes)			
	Light Shapes and American Standard	Ton	420.00	1,230.00
.5	STRUCTURAL TUBING (Steel Pipe, Square & Rectangular)	Ton	430.00	1,250.00
.6	PLATES AND BARS	Ton	450.00	1,315.00
.7	BOX GIRDERS	Ton	720.00	1,250.00
.8	TRUSSES	Ton	760.00	1,700.00
	Add for Welded Field Connections	Ton	170.00	-
	Add for High Strength Bolting	Ton	-	180.00
	Add for Riveted	Ton	320.00	-
	Add for Weathering Steel	Ton	-	250.00
	Add for Galvanizing	Ton	-	350.00

0502.0 MISCELLANEOUS METALS (L+M) (Ironworkers)

		UNIT	LABOR	MATERIAL
.1	BOLTS (with Nuts & Washers in Concrete or Masonry)	See 0201.36 for Hole Drilling in Field		
	1/2" x 4"	Each	2.00	1.50
	5/8" x 4"	Each	2.10	1.75
	3/4" x 4"	Each	2.30	2.25
	1" x 4"	Each	2.45	3.00
.2	CASTINGS			
	Manhole Ring & Cover -24" Light Duty	Each	74.00	250.00
	24" Heavy Duty	Each	84.80	300.00
	Wheel Guards- 2' 6"	Each	79.25	190.00
.3	CORNER GUARDS- 3" x 3" x 1/4" Steel	LnFt	4.90	6.00
	4" x 4" x 1/4" Steel	LnFt	5.50	7.50
	Galv.	LnFt	5.00	9.00
.4	FIRE ESCAPES- 24" Steel Grate	Per Floor	935.00	2,600.00
		Tread	58.00	160.00
.5	FRAMES, CHANNEL- 6" - 8.2 lbs	LnFt	4.45	9.90
	8" - 11.5 lbs	LnFt	5.85	13.40
	10" - 15.3 lbs	LnFt	7.90	24.00
.6	GRATINGS- Steel - 1" x 1/8"	SqFt	2.10	7.10
	Aluminum - 1" x 1.8"	SqFt	2.10	14.40
	Cast Iron #2 - Light Duty	SqFt	2.10	9.70
	Heavy Duty	SqFt	1.80	18.50
.7	GRILLES	SqFt	3.20	8.30
.8	HANDRAILS - 1 1/2"			
	Wall Mounted - Single Rail- Steel	LnFt	5.85	9.80
	Aluminum	LnFt	6.90	36.50
	Floor Mounted - 2 Rail- Steel	LnFt	7.90	21.50
	3 Rail - Steel	LnFt	8.70	26.80
	Add for Galvanizing	LnFt	-	20%
.9	LADDERS - Steel	LnFt	10.60	34.00
	Aluminum	LnFt	10.30	47.50
.10	LINTELS, LOOSE	Ton	490.00	1,060.00
	3" x 3" x 1/4" - 4.9 lbs per foot	LnFt	1.65	3.70
	4" x 3" x 1/4" - 5.8 lbs per foot	LnFt	2.00	4.20
	4" x 3 1/2" x 1/4" - 6.2 lbs per foot	LnFt	2.15	4.50
	4" x X3 1/2" x 5/16" - 7.7 lbs per foot	LnFt	2.70	5.30
	4" x X3 1/2" x 3/8" - 9.1 lbs per foot	LnFt	3.15	6.20
	4" x 4" x 1/4" - 6.6 lbs per foot	LnFt	2.55	4.90
	Add for Punching Holes	Each	-	3.45
	Add for Galvanizing	LnFt	-	1.05

DIVISION #5 - METALS

			UNIT	LABOR	MATERIAL (Subcontracted)
0502.0	**MISCELLANEOUS METALS (Cont'd...)**				
.11	PIPE GUARDS - 4" - 4' Above Grade &		LnFt	9.10	6.60
	(Bollards in Concrete) - 4' Below Grade		LnFt	9.60	9.00
.12	PLATE, CHECKERED - 1/4" - 10.2 lbs		SqFt	1.60	6.10
	1/8" - 6.2 lbs		SqFt	1.30	5.15
.13	STAIRS, METAL PAN - 4' Wide x 10' (No Concrete)		PerFl	825.00	3,100.00
	Including Supports		Tread	58.00	200.00
.14	STAIRS, CIRCULAR - 3' x 10' - Aluminum		PerFl	460.00	3,150.00
	Steel		PerFl	510.00	3,350.00
	Cast Iron		PerFl	550.00	3,770.00
.15	WIRE GUARDS		SqFt	2.20	8.10
.16	TRENCH FRAME AND COVER - 2'		LnFt	5.70	9.50
0503.0	**ORNAMENTAL METALS (M) (Ironworkers)**				
	Non Ferrous - See 0502.0				
0503.0	**OPEN WEB JOISTS (L+M) (Ironworkers)**				
.1	STANDARD BAR JOISTS (Spans to 48')				
	Small Jobs (to 20 ton) - Spans to 15'		Ton	330.00	820.00
	25'		Ton	335.00	800.00
	35'		Ton	320.00	770.00
	Medium Jobs (20 - 50 ton) - Spans to 15'		Ton	320.00	820.00
	25'		Ton	315.00	790.00
	35'		Ton	310.00	770.00
	48'		Ton	300.00	740.00
	Large Jobs (50 ton & up) - Spans to 15'		Ton	310.00	810.00
	25'		Ton	300.00	790.00
	35'		Ton	300.00	770.00
	48'		Ton	290.00	760.00
.2	LONG SPAN JOISTS (Spans 48' to 150')				
	Small Jobs (to 30 ton) - Spans to 60'		Ton	330.00	810.00
	Medium Jobs (30 -100 ton) - Spans to 60'		Ton	325.00	820.00
	80'		Ton	310.00	845.00
	100'		Ton	290.00	855.00
	Large Jobs (100 ton & Up) - Spans to 60'		Ton	305.00	800.00
	80'		Ton	300.00	840.00
	100'		Ton	295.00	870.00
	120'		Ton	285.00	940.00
	150'		Ton	280.00	990.00
0505.0	**METAL DECKING (L+M) (Sheet Metal Workers)**				
.1	SHORT SPAN (Spans to 6')				
	(Baked Enamel - Ribbed, V Beam & Seam)				
	1 1/2" Deep - 18 Ga		Sq	49.50	120.00
	20 Ga		Sq	46.50	94.00
	22 Ga		Sq	44.40	88.00
	Add for Galvanized		Sq	-	10%
.2	LONG SPAN (Spans to 12') (Enameled & Open Bottom)				
	3" Deep - 16 Ga		Sq	64.05	376.00
	18 Ga		Sq	60.00	332.00
	20 Ga		Sq	59.00	290.00
	22 Ga		Sq	57.00	235.00
	4 1/2" Deep - 16 Ga		Sq	69.20	445.00
	18 Ga		Sq	68.20	400.00
	20 Ga		Sq	66.10	365.00
	Add for Acoustical		Sq	-	20%
	Add for Galvanized		Sq	-	20%

DIVISION #5 - METALS

			UNIT	LABOR	MATERIAL
				(Subcontracted)	
0505.0	**METAL DECKING (L+M) (Sheet Metal), Cont'd...**				
.3	CELLULAR - Electric and Air Flow				
	3" Deep - 16 - 16 Ga Galvanized		Sq	77.00	550.00
	18 - 18 Ga Galvanized		Sq	72.00	515.00
	4 1/2" Deep- 16 - 16 Ga Galvanized		Sq	90.00	680.00
	18 - 18 Ga Galvanized		Sq	84.00	575.00
.4	CORRUGATED METAL				
	Standard	Spans to Approx.			
	Black - .015"	3' 6"	Sq	37.00	57.00
	Galvanized - .015"	3' 6"	Sq	37.00	70.00
	Heavy Duty				
	Black - 26 Ga	5' 0"	Sq	40.00	67.00
	Galvanized - 26 Ga	5' 0"	Sq	40.00	82.00
	Super Duty				
	Black - 24 Ga	7' 0"	Sq	42.00	95.00
	Galvanized - 24 Ga	7' 0"	Sq	42.00	105.00
	Black - 22 Ga	7' 6"	Sq	43.00	115.00
	Galvanized - 22 Ga	7' 6"	Sq	43.00	119.00
.5	BOX RIB				
	2" - 24 Ga		Sq	58.00	340.00
	22 Ga		Sq	62.00	360.00
	20 Ga		Sq	66.00	390.00
0506.0	**METAL SIDINGS & ROOFINGS (L+M) (Sheet Metal)**				
.1	INDUSTRIAL				
.11	Steel (24 Ga - See 0505 for Other Gauges)				
	Corrugated - Galvanized		Sq	72.00	100.00
	Ribbed - Enameled		Sq	66.00	96.00
	Add for Liner Panel		Sq	33.00	90.00
	Add for Insulation		Sq	28.00	28.00
.12	Aluminum- Corrugated		Sq	62.00	120.00
.2	ARCHITECTURAL (Prefinished)				
.21	Steel - Baked Enameled		Sq	95.00	160.00
	Porcelain Enameled		Sq	120.00	200.00
	Acrylic Enameled		Sq	115.00	177.00
	Plastic Faced		Sq	115.00	240.00
.22	Aluminum - Anodized		Sq	112.00	180.00
	Plastic Faced		Sq	114.00	240.00
	Porcelainized		Sq	117.00	230.00
	Add for Liner Panel and Insulation		Sq	44.00	105.00
.23	Stainless Steel		Sq	185.00	620.00
.24	Protected Metal (Asphalt on Steel w/Finish)		Sq	136.00	345.00
	See 1310 for Metal Buildings				
0507.0	**LIGHT GAGE FRAMING (L+M) (Ironworkers)**				
	3 5/8"- 16 Ga		LnFt	.59	2.85
	6"- 16 Ga		LnFt	.62	3.60
	8"- 16 Ga		LnFt	.75	4.20
	10"- 16 Ga		LnFt	.90	5.30
	Average - Wall		Lb	.35	.75
	Average - Hanging		Lb	.70	.90
0508.0	**EXPANSION CONTROL (L+M) (Ironworkers)**				
	4"- Aluminum		LnFt	6.90	36.00
	Bronze		LnFt	7.40	93.00
	Stainless Steel		LnFt	7.40	96.00
	2"- Aluminum		LnFt	5.55	27.00
	Bronze		LnFt	6.90	55.00
	Stainless Steel		LnFt	6.90	60.00

DIVISION #5 - METALS

		Sq. Ft. Cost
0501.0	**Structural Steel Frame**	
	To 30 Ton 20' Span	6.20
	24' Span	6.65
	28' Span	7.20
	32' Span	7.90
	36' Span	8.80
	40' Span	9.50
	44' Span	10.10
	48' Span	11.00
	52' Span	12.40
	56' Span	13.40
	60' Span	14.25
	Deduct for Over 30 Ton	5%
0504.0	**Open Web Joists**	
	To 20 Ton 20' Span	1.80
	24' Span	1.85
	28' Span	1.90
	32' Span	2.05
	36' Span	2.20
	40' Span	2.50
	48' Span	2.95
	52' Span	4.10
	56' Span	3.55
	60' Span	4.20
	Deduct for Over 20 Ton	10%
0505.0	**Metal Decking**	
	1 1/2" Deep - Ribbed - Baked Enamel - 18 Ga	2.10
	20 Ga	2.00
	22 Ga	1.75
	3" Deep - Ribbed - Baked Enamel - 18 Ga	4.55
	20 Ga	4.15
	22 Ga	3.60
	4 1/2" Deep - Ribbed - Baked Enamel - 16 Ga	6.28
	18 Ga	5.30
	20 Ga	4.65
	3" Deep - Cellular - 18 Ga	7.50
	16 Ga	8.40
	4 1/2" Deep - Cellular - 18 Ga	8.55
	16 Ga	10.10
	Add for Galvanized	10%
	Corrugated Black Standard .015	1.10
	Heavy Duty - 26 Ga	1.20
	S. Duty - 24 Ga	1.70
	22 Ga	1.75
	Add for Galvanized	15%
0506.0	**Metal Sidings and Roofings**	
	Aluminum- Anodized	3.40
	Porcelainized	7.10
	Corrugated	1.90
	Enamel, Baked Ribbed	3.00
	24 Ga Ribbed	2.10
	Acrylic	3.55
	Porcelain Ribbed	4.00
	Galvanized- Corrugated	1.95
	Plastic Faced	4.30
	Protected Metal	5.20
	Add for Liner Panels	1.60
	Add for Insulation	.70

DIVISION #6 - WOOD & PLASTICS

Wage Rates (Including Fringes) & Location Modifiers
July 2002-2003

	Metropolitan Area		Carpenter Rate	Wage Rate Location Modifier
1.	Akron	*	33.53	114
2.	Albany-Schenectady-Troy		29.57	102
3.	Atlanta		22.01	80
4.	Austin	**	21.99	74
5.	Baltimore		26.09	92
6.	Birmingham	**	21.98	75
7.	Boston		39.38	136
8.	Buffalo-Niagara Falls		38.38	128
9.	Charlotte	**	20.03	67
10.	Chicago-Gary	*	39.06	126
11.	Cincinnati		27.87	95
12.	Cleveland		33.84	117
13.	Columbus	*	28.70	99
14.	Dallas-Fort Worth	**	19.96	68
15.	Dayton	*	31.28	105
16.	Denver-Boulder		24.07	81
17.	Detroit		38.18	127
18.	Flint	*	30.35	100
19.	Grand Rapids		30.34	100
20.	Greensboro-West Salem	**	20.13	67
21.	Hartford-New Britain	*	34.23	116
22.	Houston	**	23.10	73
23.	Indianapolis		31.23	104
24.	Jacksonville	**	24.59	82
25.	Kansas City		29.72	98
26.	Los Angeles-Long Beach		32.99	122
27.	Louisville	**	24.37	82
28.	Memphis	**	20.73	87
29.	Miami	**	22.22	75
30.	Milwaukee	*	34.42	112
31.	Minneapolis-St. Paul		34.57	113
32.	Nashville	**	23.63	79
33.	New Orleans	**	21.59	73
34.	New York		55.66	193
35.	Norfolk-Portsmouth	**	23.23	79
36.	Oklahoma City	**	22.22	75
37.	Omaha-Council Bluffs		23.14	75
38.	Orlando	**	31.28	89
39.	Philadelphia	*	40.77	135
40.	Phoenix	**	24.78	95
41.	Pittsburgh	*	32.47	109
42.	Portland	*	33.52	113
43.	Providence-Pawtucket	*	33.63	114
44.	Richmond	**	23.19	78
45.	Rochester		34.95	116
46.	Sacramento	*	40.60	132
47.	St. Louis		35.57	117
48.	Salt Lake City-Ogden	**	22.99	70
49.	San Antonio	**	21.57	75
50.	San Diego	*	33.68	124
51.	San Francisco-Oakland-San Jose		40.60	130
52.	Seattle-Everett		34.89	116
53.	Springfield-Holyoke-Chicopee	*	32.64	110
54.	Syracuse	*	28.57	114
55.	Tampa-St. Petersburg	**	26.62	91
56.	Toledo	*	34.41	117
57.	Tulsa	**	22.65	75
58.	Tucson	**	28.09	95
59.	Washington D.C.	*	24.72	87
60.	Youngstown-Warren	*	30.39	107
	AVERAGE		**29.52**	

Impact Ratio: Labor 50%, Material 50%
* Contract Not Settled - Wage Interpolated
** Non Signatory or Open Shop Rate

DIVISION #6 - WOOD & PLASTICS

		PAGE
0601.0	**ROUGH CARPENTRY (CSI 06100)**	**6-3**
.1	LIGHT FRAMING AND SHEATHING	6-3
.11	Joists, Plates and Framing	6-3
.12	Studs, Plates and Framing	6-3
.13	Bridging	6-3
.14	Rafters	6-3
.15	Stairs	6-3
.16	Subfloor Sheathing	6-4
.17	Floor Sheathing	6-4
.18	Wall Sheathing	6-4
.19	Roof Sheathing	6-4
.2	HEAVY FRAMING AND DECKING	6-5
.21	Columns and Posts	6-5
.22	Decking	6-5
.3	MISCELLANEOUS CARPENTRY	6-6
.31	Blocking and Bucks	6-6
.32	Grounds, furring, Sleepers and nailers	6-6
.33	Cant Strips	6-6
.34	Building Papers and Sealers	6-6
0602.0	**FINISH CARPENTRY (CSI 006200)**	**6-7**
.1	FINISH SIDINGS AND FACING MATERIALS (Exterior)	6-7
.11	Boards, Beveled, and Lap Siding	6-7
.12	Plywood - Siding	6-7
.13	Hardboard - Paneling and lap Siding	6-7
.14	Shingles	6-7
.15	Facias and Trim	6-8
.16	Soffits	6-8
.2	FINISH WALLS AND CEILING MATERIALS (Interior)	6-8
.21	Boards	6-8
.22	Hardboard or Pressed Wood (4' x 8' Panels)	6-8
.23	Plywood Paneling (4' x 8' Panels and Prefinished)	6-8
.24	Gypsum Board and Metal Framing	6-9
0603.0	**MILLWORK AND CUSTOM WOODWORK**	**6-10**
.1	CUSTOM CABINET WORK	6-10
.2	COUNTER TOPS	6-10
.3	CUSTOM DOORS AND FRAMES	6-11
.4	MOULDINGS AND TRIM	6-11
.5	STAIRS	6-12
.6	SHELVING	6-12
.7	CUSTOM PANELING	6-12
0604.0	**GLUE LAMINATED (CSI 06180)**	**6-12**
.1	ARCHES AND RIGID FRAMES	6-12
.2	BEAMS, JOISTS AND PURLINS	6-12
.3	DECKING	6-12
0605.0	**PREFABRICATED WOOD AND PLYWOOD COMPONENTS (CSI 06170)**	**6-12**
.1	WOOD TRUSSED RAFTERS	6-12
.2	WOOD FLOOR AND ROOF TRUSS JOISTS	6-12
.21	Open Wood Web - Gusset Plate Connection	6-12
.22	Open Metal Web - Pin Connection	6-12
.23	Plywood Web	6-13
.3	LAMINATED VENEER STRUCTURAL BEAMS	6-13
.31	Micro Lam - Plywood	6-13
.32	Glue Lam - Dimension	6-13
0606.0	**WOOD TREATMENT (CSI 06300)**	**6-13**
0607.0	**ROUGH HARDWARE (CSI 06050)**	**6-13**
.2	NAILS, SCREWS, ETC.	6-13
.2	JOIST HANGERS	6-13
.3	BOLTS	6-13
0608.0	**EQUIPMENT**	**6-13**
6A	**QUICK ESTIMATING**	**6A-14 thru 6A-19**

DIVISION #6 - WOOD & PLASTICS

0601.0 ROUGH CARPENTRY (Carpenters)

		UNIT	LABOR	MATERIAL
.1	**LIGHT FRAMING & SHEATHING (Standard - Hem-Fir)**			
.11	Joists, Beams and Headers - 16" O.C.			
	2"x 6" Floor	BdFt	.56	.53
	2"x 8" Floor	BdFt	.54	.53
	2"x10" Floor	BdFt	.52	.54
	2"x12" Floor	BdFt	.54	.56
	Add for Ceiling Joists - 2nd Fl or Roof	BdFt	.10	-
	Add for Door, Window & Misc. Headers	BdFt	.37	-
	Add for Sloped Installations	BdFt	40%	-
	Add for Ledgers	BdFt	.25	-
.12	Studs, Plates and Miscellaneous Framing - 16" O.C.			
	2" x 2"	BdFt	.79	.56
	2" x 3"	BdFt	.74	.54
	2" x 4"	BdFt	.64	.52
	2" x 6"	BdFt	.62	.53
	Add for 2nd Floor and Above per Floor	BdFt	.10	-
	Add for Bolted Plates or Sills	BdFt	.54	-
	Add for Each 1' Above 8' in Length	BdFt	.04	-
	Add for Fire Stops, Fillers, and Nailers	BdFt	.34	-
	Add for Soffit & Suspended Framing 2"x 4"	BdFt	.79	-
	See 0602 for Metal Studs and Plates			
.13	Bridging - 16" O.C.			
	1" x 3" Wood Diagonal	Set	1.40	.38
	2" x 4" Diagonal	Set	1.60	.95
	1" Metal Diagonal - 18 Ga	Set	1.40	.95
	2" x 8" Solid	BdFt	1.65	.60
.14	Rafters and Field-Made Trussed Rafters - 16" O.C.			
	(See 0605 for Prefabrication)			
	2" x 4"	BdFt	.83	.52
	2" x 6"	BdFt	.81	.53
	2" x 8"	BdFt	.78	.54
	2" x 10"	BdFt	.73	.56
	2" x 4" Trusses - Make up in Field	BdFt	.64	.72
	Erection	BdFt	.44	-
	Add for Hips and Valleys	BdFt	.44	-
	Add for Bracing	BdFt	.84	.60
.15	Stairs			
	2" x 10" and 2" x 12" for Stringers	BdFt	1.80	.60
	1" x 8" and 2" x 10" for Risers & Stairs	BdFt	1.28	.61
	Average for Stringers, Treads and Risers	BdFt	1.53	.72
	Add to Above Framing Where Applicable:			
	Circular Installations	BdFt	60%	-
	Diagonal Installations	BdFt	10%	7%
	Kiln Dry Lumber	BdFt	-	15%
	Construction Grade Lumber	BdFt	-	.06
	Lumber Over 20'	BdFt	-	.25
	Winter Construction	BdFt	10%	-
	Add for Pressure Treating			
	2" x 4" - 2" x 6" - 2" x 8"	BdFt	-	.15
	2" x 10" - 2" x 12"	BdFt	-	.24
	Add for Fire Treating	BdFt	-	.25

DIVISION #6 - WOOD & PLASTICS

0601.0 ROUGH CARPENTRY, Cont'd...

		UNIT	LABOR	MATERIAL
.16	Sub Floor Sheathing (Structural)			
	1" x 8" and 1" x 10" #3 Pine	BdFt	.38	.56
	1/2" x 4' x 8' Plywood - CD Exterior	SqFt	.36	.41
	5/8" x 4' x 8' Plywood - CD Exterior	SqFt	.37	.49
	3/4" x 4' x 8' Plywood - CD Exterior	SqFt	.38	.60
	3/4" x 4' x 8' Plywood - T&G, CD Exterior	SqFt	.44	.84
	5/8" x 4' x 8' Particle Board	SqFt	.38	.40
	3/4" x 4' x 8' Particle Board	SqFt	.39	.48
	3/4" x 4' x 8' Particle Board, T&G	SqFt	.43	.60
.17	Floor Sheathing (Over Sub Floor) With Partitions in Place			
	1/4" x 4' x 8' Plywood - CD	SqFt	.39	.34
	3/8" x 4' x 8' Plywood - CD	SqFt	.40	.35
	1/2" x 4' x 8' Plywood - CD	SqFt	.41	.46
	5/8" x 4' x 8' Plywood - CD	SqFt	.42	.57
	3/8" x 4' x 8' Particle Board	SqFt	.40	.24
	1/2" x 4' x 8' Particle Board	SqFt	.41	.29
	5/8" x 4' x 8' Particle Board	SqFt	.42	.40
	3/4" x 4' x 8' Particle Board	SqFt	.43	.50
.18	Wall Sheathing			
	1" x 8" and 1" x 10" #3 Pine	BdFt	.43	.58
	3/8" x 4' x 8' Plywood - CD Exterior	SqFt	.42	.36
	1/2" x 4' x 8' Plywood - CD Exterior	SqFt	.43	.47
	5/8" x 4' x 8' Plywood - CD Exterior	SqFt	.44	.57
	3/4" x 4' x 8' Plywood - CD Exterior	SqFt	.46	.60
	1/2" x 4' x 8' Fiberboard Impregnated	SqFt	.41	.26
	25/32" x 4' x 8' Fiberboard Impregnated	SqFt	.42	.29
	1" x 2' x 8' T&G Styrofoam	SqFt	.44	.33
	2" x 2' x 8' T&G Styrofoam	SqFt	.58	.63
	1/2" x 4' x 8' Wafer Board	SqFt	.42	.28
	5/8" x 4' x 8' Wafer Board	SqFt	.45	.37
.19	Roof Sheathing - Flat			
	1" x 6" and 1" x 8" #3 Pine	BdFt	.43	.58
	1/2" x 4' x 8' Plywood - CD Exterior	SqFt	.40	.47
	5/8" x 4' x 8' Plywood - CD Exterior	SqFt	.42	.57
	3/4" x 4' x 8' Plywood - CD Exterior	SqFt	.43	.60
	Add for Sloped Installation	SqFt	.12	-
	Add Steep Sloped Installation (over 5-12)	SqFt	.29	-
	Add for Dormer Work	SqFt	.53	-
	Add to Above Sheathing where Applicable:			
	Fire Treating	SqFt	-	.26
	Pressure Treating	SqFt	-	.22
	Winter Construction	SqFt	10%	-
	Plywood - AC Exterior (Good One Side)	SqFt	-	.21
	Plywood - AD Exterior (Good One Side)	SqFt	-	.27
	Diagonal Installations	SqFt	8%	7%
	Plywood 10' Lengths	SqFt	-	.18

Recommended 10% Decrease on Labor Units for Residential Construction

DIVISION #6 - WOOD & PLASTICS

0601.0 ROUGH CARPENTRY, Cont'd...

.2   HEAVY FRAMING AND DECKING (Construction-Grade Fir, Air Dried S4S)
    See 0604.0 for Laminated Heavy Framing & Decking.
    See 0606.0 for Wood Treatments.
    See 0109.8 for Wood Curbs and Walls.

	UNIT	LABOR	MATERIAL
.21 Columns and Posts			
4"x 4"	MBF	535.00	920.00
4"x 6"	MBF	530.00	940.00
6"x 6"	MBF	495.00	1,130.00
8"x 8"	MBF	490.00	1,280.00
12"x12"	MBF	465.00	1,700.00
Add for Fitting at Base and Heat	MBF	160.00	-
Beams and Joists			
4" x 4"	MBF	535.00	920.00
4" x 6"	MBF	525.00	940.00
4" x 8"	MBF	520.00	1,080.00
4" x10"	MBF	500.00	1,150.00
6" x 6"	MBF	505.00	1,130.00
6" x 8"	MBF	490.00	1,170.00
6" x10"	MBF	485.00	1,230.00
8" x12"	MBF	475.00	1,430.00
Add to 0601.21 Above for:			
Rough Sawn	MBF	-	130.00
Mill Cutting to Length and Shape	MBF	-	130.00
Nails	MBF	-	60.00
Hauling & Unloading (if Quoted Cars)	MBF	-	80.00
Equipment & Operator Cost - Fork Lift	MBF	-	65.00
Equipment & Operator Cost - Crane	MBF	-	100.00
Add for Red Cedar	MBF	-	1,100.00
Add for Redwood	MBF	-	1,300.00
.22 DECKING			
Tongue & Groove			
Fir, Hemlock & Spruce, Construction Grade (SPF)			
4" x 4"	MBF	345.00	950.00
2" x 6" and 2" x 8"	MBF	320.00	720.00
3" x 6"	MBF	330.00	870.00
4" x 6"	MBF	340.00	970.00
Cedar - #3 and better			
3" x 4"	MBF	330.00	1,500.00
4" x 4" - 4" x 6" - 4" x 8" and 4" x 10"	MBF	325.00	1,650.00
White Pine			
3" x 6" and 3" x 8"	MBF	330.00	1,550.00
4" x 6" and 4" x 8"	MBF	340.00	1,700.00
Add for D Grade	MBF	-	1,050.00
2" x 6" Panelized Decking - 1 1/2" x 20"			
Premium Grade	SqFt	.70	2.00
Architectural Grade	SqFt	.70	1.90
Add for Select Grade	MBF	-	5%
Add for Hemlock	MBF	-	45.00
Add for Hip Roof	MBF	65.00	-
Add for Specified Lengths	MBF	-	32.00
Add for End Matched	MBF	-	42.00
Add for Hauling and Unloading	MBF	-	80.00
Add for Nails	MBF	-	50.00

    See 0604 for Laminated Decking
    See 0604.3 for Square Foot Deck Costs

DIVISION #6 - WOOD & PLASTICS

0601.0 ROUGH CARPENTRY, Cont'd...

.3	MISCELLANEOUS CARPENTRY			
		UNIT	LABOR	MATERIAL
.31	Blocking and Bucks (2" x 4" or 2" x 6")			
	Doors and Window Bucks			
	Erected Before Masonry	BdFt	1.32	.52
	or, Each	22.50	8.80
	Bolted to Masonry or Steel	BdFt	1.85	.52
	or, Each	31.50	8.80
	Nail Driven to Steel Studs or Masonry	BdFt	1.20	.52
	or, Each	20.40	8.80
	Roof Blocking - Pressure Treated			
	Edges - Bolted to Concrete	BdFt	1.40	.80
	Nailed to Wood	BdFt	.70	.75
	Openings - Nailed	BdFt	1.55	.75
	Add for 2" x 8" or 2" x 10"	BdFt	-	.05
	Add to 0601.31 for Bolts - 4' O.C.	BdFt	-	.30
.32	Grounds, Furring, Sleepers & Nailers #3 Pine			
	1" x 4" Fastened to Wood	BdFt	.90	.54
	or, LnFt	.30	.18
	1" x 3" Fastened to Wood	LnFt	.42	.15
	1" x 2" Fastened to Wood	LnFt	.40	.13
	1" x 3" Fastened to Concrete/ Masonry			
	Nailed	LnFt	.57	.30
	Gun Driven (Incl Fasteners)	LnFt	.50	.35
	Pre Clipped	LnFt	.72	.32
	Add for Ceiling Work	LnFt	.16	-
	2" x 4" Not Suspended Framing	BdFt	1.37	.52
	Suspended Framing	BdFt	1.70	.52
	2" x 2" Not Suspended Framing	BdFt	1.75	.56
	Suspended Framing	BdFt	2.50	.56
.33	Cant Strips			
	Pressure Treated			
	Cut from 4" x 4" - At Roof Edges	BdFt	.43	.45
	or, LnFt	.58	.60
	4" x 4" - At Roof Openings	BdFt	.62	.45
	or, LnFt	.82	.60
	6" x 6" - At Roof Edges	BdFt	1.51	2.10
	or, LnFt	1.00	1.40
	6" x 6" - At Roof Openings	BdFt	2.10	2.10
	or, LnFt	1.40	1.40
	Add for Bolting (Bolts - 4' O.C.)	BdFt	.63	.54
	Not Treated			
	Cut from 4" x 4" - At Roof Edges	BdFt	.42	.38
	or, LnFt	.58	.50
	6" x 6" - At Roof Edges	BdFt	1.52	1.75
	or, LnFt	1.00	1.15
.34	Building Papers and Sealers			
	15" Felt (432 Sq.Ft. to Roll)	SqFt	.07	.03
	20" Felt (432 Sq.Ft. to Roll)	SqFt	.08	.04
	30" Felt (216 Sq.Ft. to Roll)	SqFt	.09	.06
	Cotton and Glass Asphalt Saturated	SqFt	.08	.06
	Polyethylene 4 mil	SqFt	.07	.02
	6 mil	SqFt	.08	.03
	30# Red Rosin Paper	SqFt	.08	.13
	Sill Sealer 6" x 100'	SqFt	.24	.21

DIVISION #6 - WOOD & PLASTICS

0602.0 FINISH CARPENTRY (Carpenters)

.1 FINISH SIDINGS AND FACING MATERIAL (Exterior)
(See Quick Estimating for Square Foot Costs)

	UNIT	LABOR	MATERIAL
.11 Boards, Beveled and Lap Siding			
Cedar - Beveled - Clear 1/2" x 4"	BdFt	1.00	1.50
1/2" x 6"	BdFt	.94	1.61
Beveled - Rough Sawn 7/8" x 8"	BdFt	.87	1.40
7/8" x 10"	BdFt	.83	1.60
7/8" x 12"	BdFt	.81	1.65
Board - Rough Sawn 3/4" x 12"	BdFt	.79	1.60
Redwood Beveled - Clear Heart 1/2" x 4"	BdFt	1.00	1.65
1/2" x 6"	BdFt	.93	1.82
1/2" x 8"	BdFt	.87	1.93
5/8" x 10"	BdFt	.81	2.63
3/4" x 6"	BdFt	.85	2.45
3/4" x 8"	BdFt	.81	2.55
Board - Clear Heart 3/4" x 6"	BdFt	.78	3.05
3/4" x 8"	BdFt	.75	4.15
Board, - T&G 1" x 4" & 6"	BdFt	.81	3.30
Rustic Beveled - 1" x 8" & 10"	BdFt	.78	3.50
5/4" x 6" & 8"	BdFt	.75	3.40
Add for Metal Corners	Each	.82	.33
Add for Mitering Corners	Each	2.65	-
.12 Plywood - Siding			
Fir - AC - Smooth One Side 1/4"	SqFt	.53	.49
3/8"	SqFt	.55	.53
Fir - Rough Faced 3/8"	SqFt	.57	.60
Fir - AC - Smooth One Side 1/2"	SqFt	.58	.60
5/8"	SqFt	.61	.75
3/4"	SqFt	.65	.83
Fir - Grooved and Rough Faced 5/8"	SqFt	.63	.82
Cedar - Rough Sawn 3/8"	SqFt	.56	.94
5/8"	SqFt	.63	1.22
3/4"	SqFt	.66	1.38
Cedar - Grooved & Rough Faced (4" & 8") 5/8"	SqFt	.63	1.35
Add for 4' x 10' and 4' x 9' Sheets	SqFt	-	.14
Add for Wood Batten Strips 1" x 2" Pine	LnFt	.55	.30
Add for Horizontal Joint Flashing (Spline)	LnFt	.44	.38
Add for Medium Density Fir	SqFt	.02	.16
.13 Hardboard - Paneling and Lap Siding (Primed)			
7/16" - Rough Textured 4' x 8' Paneling	SqFt	.48	.95
7/16" - Grooved 4' x 8' Paneling	SqFt	.53	.95
7/16" - Stucco Board 4' x 8' Paneling	SqFt	.54	1.00
Lap Siding - 7/16" x 8"	BdFt	.72	.68
7/16" x 12"	BdFt	.79	.73
Add for Prefinishing	BdFt	-	.15
.14 Shingles - Wood			
16" Red Cedar - 12" to Weather - #1	SqFt	.75	1.80
#2	SqFt	.72	1.35
#3	SqFt	.65	1.00
Fancy Butt Red Cedar 7" #1	SqFt	.77	3.75
Red Cedar Hand Split - #1 1/2" to 3/4"	SqFt	.90	1.40
3/4" to 1 1/4" - #1	SqFt	.93	1.68
3/4" to 1 1/4" - #2	SqFt	.90	1.78
3/4" to 1 1/4" - #3	SqFt	.88	1.68
Add for Fire Retardant	SqFt	-	.42
Add for 3/8" Backer Board	SqFt	.28	.22
Add for Metal Corners	Each	.75	.33
Add for Ridges, Hips and Corners	LnFt	1.27	.85
Add for 8" to Weather	SqFt	25%	-

See Division 0708 for Roof Shingles

DIVISION #6 - WOOD & PLASTICS

0602.0 FINISH CARPENTRY, Cont'd...

					MATERIAL		
				Pine	Cedar	Redwood	Plywood
		UNIT	LABOR	#2	#3	Clear	5/8" ACX
.15	Facias and Trim						
	All Sizes (average)	BdFt	.74	.75	1.60	3.60	.75
	1" x 3"	LnFt	.46	.28	.40	.85	.19
	1" x 4"	LnFt	.53	.33	.48	1.24	.25
	1" x 6"	LnFt	.62	.45	.85	1.75	.38
	1" x 8"	LnFt	.67	.75	1.18	2.15	.50
	1" x 12"	LnFt	.73	.75	1.60	3.60	.75
	Add for Clear Grades			400%	100%	-	30%

		UNIT	LABOR	MATERIAL
.16	Soffits			
	1/2" Plywood Fir ACV	SqFt	.90	.60
	Aluminum	SqFt	.90	1.20
	Plastic - Egg Crate	SqFt	.90	1.30
.2	FINISH WALLS AND CEILING MATERIALS (Interior)			
.21	Boards (See Quick Est. - Size/Cutting Allowance Added)			
	Cedar - #3 Grade - 1" x 6"	SqFt	.76	1.85
	Knotty 1" x 6"	SqFt	.74	1.95
	1" x 8"	SqFt	.71	2.15
	D Grade 1" x 4"	SqFt	.76	2.70
	1" x 6"	SqFt	.74	2.80
	1" x 8"	SqFt	.72	2.70
	Aromatic 1" x 6"	SqFt	.76	2.80
	Redwood - Constr. 1" x 6"	SqFt	.80	2.05
	1" x 8"	SqFt	.74	2.25
	Add for Clear	SqFt	10%	100%
	Fir - Beaded 5/8" x 4"	SqFt	.80	1.85
	#2 Grade 1" x 6"	SqFt	.75	2.05
	1" x 8"	SqFt	.72	2.00
	1" x 10"	SqFt	.68	2.05
.22	Hardboard & Pressed Woods (4' x 8' Panels)			
	Tempered 1/8"	SqFt	.49	.30
	1/4"	SqFt	.54	.33
	Pegboard (Perforated) 1/8"	SqFt	.50	.36
	1/4"	SqFt	.55	.48
	Plastic Faced 1/4"	SqFt	.58	.60
	Flake Board 1/8"	SqFt	.45	.30
	Add for Metal or Plastic Mouldings	LnFt	.47	.35
	Add for Pre-finished	SqFt	-	.16
.23	Plywood Paneling (4' x 8' Panels, Prefinished)			
	Birch, Natural 1/4"	SqFt	.78	1.00
	3/4"	SqFt	.89	1.60
	White Select 1/4"	SqFt	.79	1.50
	Oak, Red - Rotary Cut 1/4"	SqFt	.79	.92
	3/4"	SqFt	.90	1.95
	White 1/4"	SqFt	.78	1.95
	Cedar, Aromatic 1/4"	SqFt	.79	1.60
	Cherry 3/4"	SqFt	.90	2.25
	Pecan 3/4"	SqFt	.90	2.25
	Walnut 1/4"	SqFt	.78	2.25
	3/4"	SqFt	.90	3.40
	Mahogany, Lauan 1/4"	SqFt	.78	.65
	3/4"	SqFt	.90	1.10
	African 3/4"	SqFt	.90	2.35
	Knotty Pine 1/4"	SqFt	.78	1.10
	3/4"	SqFt	.90	1.75
	Slat Wall - Painted 3/4"	SqFt	.90	1.45
	Oak 3/4"	SqFt	.95	2.25
	Add for V Groove and T&G	SqFt	-	.26
	Add for Glued-On Applications	SqFt	.15	.08
	Add for 10' Lengths	SqFt	.10	.22

DIVISION #6 - WOOD & PLASTICS

0602.0 FINISH CARPENTRY, Cont'd...

.24 Gypsum Board and Framing
(See 0902 - Subcontracted Work Costs)
(See 0902 - Metal Framing SqFt Costs)

	UNIT	LABOR	MATERIAL
Finish Board (Walls to 8' - Screwed On)			
3/8" 4' x 8' and 12'	SqFt	.31	.21
1/2" 4' x 8' and 12'	SqFt	.32	.23
5/8" 4' x 8' and 12	SqFt	.33	.26
1" 4' x 8' and 12' Core Board	SqFt	.34	.48
1" 4' x 8' and 12' Bead Board	SqFt	.32	.48
Add for Adhesive Application	SqFt	.07	.02
Deduct for Nail-on Application	SqFt	.02	-
Add for: Fire Resistant Board	SqFt	.02	.07
Add for: Foil Backed (Insulating)	SqFt	.02	.10
Add for: Moisture Resistant Type 5/8" & 1/2"	SqFt	.02	.06
Add for: Work Above 8'	SqFt	.06	-
Add for: Work Over 2-Story per Story	SqFt	.06	-
Add for: Ceiling Work - To Wood	SqFt	.12	-
To Steel	SqFt	.31	-
Add for: Resilient Clip Application	SqFt	.30	.06
Add for: Small Cut Up Areas	SqFt	.33	.04
Add for: Beam and Column Work	SqFt	.42	.04
Add for: Circular Construction	SqFt	.36	.05
Add for: Vinyl Faced Board			
3/8" x 4' x 8' - Cost Variable	SqFt	.36	.70
1/2" x 4' x 8' - Cost Variable	SqFt	.40	.75
Metal Framing (Studs, Runners and Channels)			
Studs 25-Gauge 1 5/8"	LnFt	.46	.21
2 1/2"	LnFt	.47	.22
3 5/8"	LnFt	.53	.24
6"	LnFt	.49	.33
8"	LnFt	.53	.35
Runners or Track 1 5/8"	LnFt	.44	.20
2 1/2"	LnFt	.46	.21
3 5/8"	LnFt	.48	.23
6"	LnFt	.49	.30
8"	LnFt	.53	.31
7/8" Furring Channels	LnFt	.46	.23
1 1/2" Furring Channels (Cold Rolled)	LnFt	.47	.30
Resilient Channels	LnFt	.44	.23
Add for Over 10' Lengths	LnFt	.16	.07
Add for 20-Gauge 6" Material	LnFt	.15	.40
Add for 16-Gauge 8" Material	LnFt	.26	.90
Metal Trim			
Casing Bead	LnFt	.63	.15
Corner Bead	LnFt	.48	.09
J Bead 3/8" - 1/2" - 5/8"	LnFt	.48	.16
L Bead 1/2" - 5/8"	LnFt	.48	.16
Expansion Bead - Metal	LnFt	.48	.50
Plastic	LnFt	.48	.25
Metal Mouldings	LnFt	.73	.40
Taping and Sanding	SqFt	.27	.06
Texturing	SqFt	.29	.08
Thin Coat Plaster - 1 Coat	SqFt	.78	.26
Sound Deadening Insulation - 2 1/2"	SqFt	.28	.24
3 1/2"	SqFt	.32	.24
6"	SqFt	.34	.31

Recommend 10% decrease on Labor Units for Residential Construction

DIVISION #6 - WOOD & PLASTICS

0603.0 MILLWORK & CUSTOM WOODWORK (M) (Carpenters) — MATERIAL

.1 CUSTOM CABINET WORK	UNIT	LABOR	Prefinished Red Oak or Birch	Plastic Laminate
See Division 11 - Stock Cabinets (Medium etc)				
Base Cabinets 35" High x 24" Deep -				
18" W	LnFt	16.00	108.00	105.00
24" W	LnFt	16.00	103.00	100.00
30" W	LnFt	15.00	98.00	95.00
36" W	LnFt	15.00	92.00	90.00
36" Wide Sink Fronts	LnFt	14.00	67.00	63.00
42"	LnFt	14.00	72.00	68.00
36" Wide Corner Cabinet	LnFt	15.00	113.00	115.00
Add for Lazy Susan	Each	20.00	108.00	105.00
Add per Drawer	Each	3.20	22.00	22.00
Upper Cabinets - 12" Deep				
30" High 18" Wide	LnFt	16.00	75.50	74.00
24" W	LnFt	16.00	70.50	69.00
30" W	LnFt	15.00	68.00	67.00
36" W	LnFt	15.00	65.00	64.00
24" High 18" Wide	LnFt	15.00	65.00	64.00
24" W	LnFt	15.00	59.75	57.50
30" W	LnFt	15.00	56.00	56.50
36" W	LnFt	14.00	56.00	55.50
15" High 18" Wide	LnFt	14.00	59.50	58.00
24" W	LnFt	14.00	56.80	56.00
30" W	LnFt	13.00	54.00	53.00
36" W	LnFt	13.00	54.00	52.50
Utility Cabinets - 84" H x 24" D				
18" Wide	LnFt	19.00	253.00	136.00
24" W	LnFt	19.00	243.00	132.00
Add - Prefinished Interior	-	-	8%	-
Deduct - Prefinished to Unfinished	-	-	12%	-
Add for Plastic Lam. Interior	-	-	-	100%
Add to Above for Special Hdwe.	Unit	7.00	-	20.60
Add - White/Red Birch/White Oak	-	-	-	20%
Add - Walnut	-	-	-	100%
China or Corner Cabinet 84"H x 36"W	Each	100.00	540.00	580.00
Oven Cabinet 84" High x 27" High	Each	95.00	432.00	455.00
Vanity Cabinets 30" High x 21" Deep				
30" Wide	LnFt	30.00	91.50	93.00
36" W	LnFt	31.00	92.00	90.00
48" W	LnFt	34.00	86.00	87.00

.2 COUNTER TOPS, 25" with Back Splash 4"	UNIT	LABOR	MATERIAL
Plastic Laminated 1 1/2"	LnFt	9.50	26.50
Deduct for No Back Splash	LnFt	2.10	5.25
Plastic Laminated Vanity Top	LnFt	8.50	32.00
Marble	LnFt	9.00	65.50
Artificial	LnFt	9.00	50.00
Wood Cutting Block, 1 1/2"	LnFt	9.00	63.00
Stainless Steel	LnFt	10.50	95.00
Polyester-Acrylic Solid Surface	LnFt	9.50	115.00
Granite - Artificial - 1 1/4"	LnFt	11.50	58.00
3/4"	LnFt	10.50	45.00
Quartz	LnFt	10.50	50.00
Plastic (Polymer)	LnFt	10.00	35.00

DIVISION #6 - WOOD & PLASTICS

0603.0 MILLWORK & CUSTOM WOODWORK, Cont'd...

.3 CUSTOM DOOR FRAMES & TRIM

	Unit	Labor	Birch	Fir	Poplar	Oak	Pine	Maple
Custom Jamb & Stop								
4 5/8" x 2'6" x 3/4" x 6'8"	Each	$37	$110	$58	$57	$102	$102	-
x 2'8" x 3/4"	Each	38	112	60	58	104	102	-
x 3'0" x 3/4"	Each	39	108	64	59	105	107	-
5 1/4" x 2'6" x 3/4" x 6'8"	Each	38	119	68	59	109	104	-
x 2'8" x 3/4"	Each	39	121	70	61	112	105	-
x 3'0" x 3/4"	Each	40	125	73	63	125	112	-
6 3/4" x 2'6" x 3/4" x 6'8"	Each	40	163	-	-	165	124	-
x 2'8" x 3/4"	Each	41	168	-	-	167	125	-
x 3'0" x 3/4"	Each	42	171	-	-	170	128	-
Stock Jamb & Stop								
4 5/8" x 2'6" x 3/4" x 6'8"	Each	35	97	33	-	96	61	-
x 2'8" x 3/4"	Each	36	99	34	-	98	63	-
x 3'0" x 3/4"	Each	35	101	35	-	100	64	-
5 1/4" x 2'6" x 3/4" x 6'8"	Each	36	107	38	46	107	65	-
x 2'8" x 3/4"	Each	37	109	39	47	109	66	-
x 3'0" x 3/4"	Each	38	113	40	48	118	68	-
6 3/4" x 2'6" x 3/4" x 6'8"	Each	38	135	-	-	150	92	-
x 2'8" x 3/4"	Each	39	137	-	-	152	94	-
x 3'0" x 3/4"	Each	40	140	-	-	154	97	-
Fire Rated Jamb & Stop								
4 5/8" x 3'0" x 1 1/16"	Each	39	171	-	-	160	-	-
5 1/4" x 3'0" x 1 1/16"	Each	40	182	-	-	170	-	-
6 3/4" x 3'0" x 1 1/16"	Each	42	187	-	-	175	-	-
Trim Only -11/16" x 2 1/4"	Each	36	104	27	37	90	73	-
(2 sides) - 3/4" x 3 1/2"	Each	42	170	33	51	118	95	-

.4 MOULDINGS & TRIM-STOCK

	Unit	Labor	Birch	Fir	Poplar	Oak	Pine	Maple
Apron 7/16" x 2"	LnFt	.79	1.23	-	.43	.91	.63	-
11/16" x 2 1/2"	LnFt	.95	1.75	-	-	1.44	1.18	1.29
Astragal 1 3/4" x 2 1/4"	LnFt	1.00	-	-	-	5.00	4.20	-
Base 7/16" x 2 3/4"	LnFt	.79	1.75	.61	.70	1.23	1.00	1.30
9/16" x 3 1/4"	LnFt	.84	2.88	-	.97	2.24	1.28	1.40
Base Shoe 7/16" x 2 3/4"	LnFt	.79	-	-	.41	.70	.44	.70
Batten Strip 5/8" x 1 5/8"	LnFt	.69	-	-	-	1.08	.87	-
Brick Mould 1 1/4" x 3"	LnFt	1.05	-	.99	-	-	1.38	-
Casing 11/16" x 2 1/4"	LnFt	.84	1.40	.72	.65	1.03	1.02	.98
3/4" x 3 1/2"	LnFt	.90	2.40	.82	.70	2.00	1.80	-
Chair Rail 5/8" x 1 3/4"	LnFt	.72	1.75	-	1.10	1.65	1.33	-
or Dado 11/16" x 2 1/4"	LnFt	.77	2.33	-	-	2.05	1.35	-
Closet Rod 1 5/16"	LnFt	1.15	-	.82	-	1.22	-	-
Corner Bead 3/4" x 3/4"	LnFt	.69	-	-	.60	1.05	.80	1.03
1 1/8" x 1 1/8"	LnFt	.76	-	-	1.05	1.55	1.38	1.45
Cove Mould 3/4" x 7/8"	LnFt	1.05	1.12	-	.50	.80	.63	.88
Crown & Cove 9/16" x 2 3/4"	LnFt	1.10	2.75	-	1.00	1.75	1.55	1.90
9/16" x 3 5/8"	LnFt	1.18	3.40	-	-	2.90	2.05	-
11/16" x 4 5/8"	LnFt	1.25	-	-	-	3.40	3.10	-
Drip Cap 11/16" x 1 3/4"	LnFt	.86	-	-	-	-	1.38	-
Half Round 1/2" x 1"	LnFt	.98	-	-	-	-	.98	-
Hand Rail 1 5/8" x 1 3/4"	LnFt	1.37	-	1.33	-	-	1.48	-
Hook Strip 5/8" x 2 1/2"	LnFt	.82	-	.82	.86	-	.96	-
Picture Mould 3/4" x 1 1/2"	LnFt	.82	.86	-	-	.98	.96	-
Quarter Round 3/4" x 3/4"	LnFt	.69	-	.37	.50	.80	.50	.80
1/2" x 1/2"	LnFt	.64	-	.28	-	-	.34	-
Sill Casement 3/4" x 1 7/8"	LnFt	1.10	-	-	-	-	1.38	-
Stool 11/16" x 2 1/2"	LnFt	1.10	3.20	1.05	1.28	2.75	1.75	-

Above prices based on 200 LnFt. Deduct for Over 200 LnFt - 20%.
Add for Walnut Trim - 100% Add for Custom Trim - 50%

DIVISION #6 - WOOD & PLASTICS

0603.0 MILLWORK & CUSTOM WOODWORK, Cont'd...

		Unit	Labor	Birch	Fir	Lauan	Oak	Pine	Walnut
.5	STAIRS								
	Treads 1 1/6" x 10 1/4"	LnFt	1.70	-	-	-	4.15	3.30	-
	Risers 3/4" x 7 1/2"	LnFt	1.50	-	-	-	6.40	2.30	-
	Skirt Bds 9/16" x 9 1/2"	LnFt	1.90	-	-	-	5.90	2.40	-
	Nosings 1/8" x 3 1/2"	LnFt	1.45	-	-	-	2.95	1.60	-
.6	SHELVING								
	12" Deep	SqFt	4.25	30.00	-	-	23.50	11.50	68.00
	8"	SqFt	5.25	25.00	-	-	21.50	10.50	56.00
.7	CUSTOM PANELING	SqFt	2.60	17.50	-	-	13.00	-	20.00
.8	THRESHOLDS 3/4" x 3 1/2"								
	Standard	LnFt	6.30	-	-	-	5.20	-	-
	With Vinyl	LnFt	6.30	-	-	-	5.72	-	-

0604.0 GLUE LAMINATED (M) (Carpenters) (F.O.B. Cars)

		UNIT	LABOR	MATERIAL
.1	ARCHES AND RIGID FRAMES: Arches	MBF	560.00	1,650.00
	Rigid Frames	MBF	570.00	1,800.00
.2	BEAMS, JOISTS & PURLINS: 4" x 6", 8" & 10"	MBF	570.00	2,170.00
	6" x 6", 8" & 10"	MBF	550.00	2,060.00
	8" x 6", 8" & 10"	MBF	540.00	1,900.00
.3	DECKS: Doug.Fir - 3" x 6" (2 1/4" x 5 1/4")	SqFt	1.65	4.55
	4" x 6" (3" x 5 1/4")	SqFt	1.75	5.15
	Cedar - 3" x 6" (2 1/4" x 5 1/4")	SqFt	1.80	5.50
	4" x 6"	SqFt	1.85	5.90
	Pine - 3" x 6" (2 1/4" x 5 1/4")	SqFt	1.70	3.85

0605.0 PREFABRICATED WOOD & PLYWOOD COMPONENTS (See 0601.22, Job Fabricated)

.1	WOOD TRUSSED RAFTERS			
	4-12 Pitch 24" O.C. With 2' Overhang 57 lb. Loading			
	16' Span	Each	29.00	60.00
	20'	Each	31.00	62.00
	22'	Each	32.00	64.00
	24'	Each	34.00	67.00
	26'	Each	36.00	74.00
	28'	Each	40.00	87.00
	30'	Each	43.00	100.00
	32'	Each	48.00	110.00
	Add for 5-12 Pitch	Each	5%	15%
	Add for 6-12 Pitch	Each	12%	25%
	Add for Gable End Type (Flying)	Each	18.00	32.00
	Add for 16" O.C.	Each	-	10%
	Add for Hip Ends - 24' Span	Each	140.00	360.00
	Add for Energy Type	Each	-	10%
.2	WOOD FLOOR AND ROOF TRUSS JOISTS			
.21	Open Wood Web - Gusset Plate Connection			
	24" O.C. Up to 23' Span x 12" - Single Chord	LnFt	1.32	4.55
	to 24' Span x 15"	LnFt	1.37	4.65
	to 27' Span x 18"	LnFt	1.42	4.75
	to 30' Span x 21"	LnFt	1.58	4.95
	to 25' Span x 15" - Double Chord	LnFt	1.47	4.20
	to 30' Span x 18"	LnFt	1.75	4.30
	to 35' Span x 21"	LnFt	2.05	4.50
	Open Metal Web - Pin Connection			
	to 25' Span x 18"	LnFt	1.40	4.30
	to 30' Span x 22"	LnFt	1.50	4.40
	to 35' Span x 30"	LnFt	1.70	4.60
	to 40' Span x 30"	LnFt	2.10	4.75

DIVISION #6 - WOOD & PLASTICS

0605.0 PREFABRICATED WOOD & PLYWOOD COMPONENTS, Cont'd...

		UNIT	LABOR	MATERIAL
.22	Plywood Web			
	24" O.C. Up to 15' Span x 9 1/2"	LnFt	1.12	2.50
	to 19' Span x 11 7/8"	LnFt	1.18	2.60
	to 21' Span x 14"	LnFt	1.32	2.95
	to 22' Span x 16"	LnFt	1.49	3.20
.3	LAMINATED VENEER STRUCTURAL BEAMS			
.31	Micro Lam (Plywood)			
	9 1/2" x 1 3/4"	LnFt	1.20	3.80
	11 7/8" x 1 3/4"	LnFt	1.27	4.30
	14" x 1 3/4"	LnFt	1.33	5.25
	16" x 1 3/4"	LnFt	1.43	5.90
	18" x 1 3/4"	LnFt	1.60	6.50
.32	Glue Lam (Dimension Lumber)			
	9" x 3 1/2" Industrial	LnFt	1.20	8.80
	12" x 3 1/2"	LnFt	1.27	10.50
	15" x 3 1/2"	LnFt	1.32	13.00
	9" x 5 1/8"	LnFt	1.20	12.20
	12" x 5 1/8"	LnFt	1.37	15.20
	15" x 5 1/8"	LnFt	1.52	18.90
	18" x 5 1/8"	LnFt	1.68	23.20
	24" x 5 1/8"	LnFt	2.00	29.20
	Add for Architectural Grade	LnFt	.22	10%
	Add for 6 3/4"	LnFt	.22	4.60
.4	DECKS - 4' x 30' w/Plywood 2 Sides - Insulated	SqFt	1.00	5.60

0606.0 WOOD TREATMENTS (Preservatives)

		UNIT	LABOR	MATERIAL
	SALTS - PRESSURE (Water Borne)			
	Dimension 2 x 4-2 x 6 and 2 x 8, 2 x 10-2 x 12	BdFt	-	.15
	2 x 10 and 2 x 12	BdFt	-	.25
	Plywood 5/8" CDX	SqFt	-	.20
	SALTS - NON PRESSURE	BdFt	-	.10
	CREOSOTE (Penta)	BdFt	-	.30
	FIRE RETARDANTS - PRESSURE			
	Dimension and Timbers	BdFt	-	.26
	Plywood 5/8" CDX	SqFt	-	.26

0607.0 ROUGH HARDWARE

		UNIT	LABOR	MATERIAL
.1	NAILS, BOLTS, ETC. Job Average	SqFt & BdFt	-	.02
	Common (50# Box) 8 Penny	CWT	-	49.00
	16 Penny	CWT	-	46.00
	Finish	CWT	-	56.00
	Add for Coated	CWT	-	60%
.2	JOIST HANGERS	Each	1.30	.93
.3	BOLTS - 5/8" x 12"	Each	1.60	1.30

0608.0 EQUIPMENT (Saws, Woodworking, Drills, Etc.)

		UNIT	LABOR	MATERIAL
	Average Job - As a Percentage of Labor	-	-	3%
	Heavy Equipment in Division 2			
	See 1-6A and 1-7A for Rental Rates			

DIVISION #6 - WOOD & PLASTICS - QUICK ESTIMATING

The costs below are average, priced as total contractor/subcontractor costs with an overhead of 5% and fee of 10% included. Units include fasteners and equipment, 35% taxes and insurance on labor, and 10% General Conditions and Equipment. Also included are nails, cutting, size and lap allowances. See Example on Page 4.

0601.0 ROUGH CARPENTRY

		UNIT	COST
.1	LIGHT FRAMING AND SHEATHING		
.11	Joists and Headers - Floor Area - 16" O.C.		
	2" x 6" Joists - with Headers & Bridging	SqFt	2.05
	2" x 8" Joists	SqFt	2.55
	2" x 10" Joists	SqFt	2.90
	2" x 12" Joists	SqFt	3.60
	Add for Ceiling Joists, 2nd Floor and Above	SqFt	8%
	Add for Sloped Installation	SqFt	15%
.12	Studs, Plates and Framing - 8' Wall Height - 16" O.C.		
	2" x 3" Stud Wall - Non-Bearing (Single Top Plate)	SqFt	1.25
	2" x 4" Stud Wall - Bearing (Double Top Plate)	SqFt	1.40
	2" x 4" Stud Wall - Non-Bearing (Single Top Plate)	SqFt	1.35
	2" x 6" Stud Wall - Bearing (Double Top Plate)	SqFt	1.80
	2" x 6" Stud Wall - Non-Bearing (Single Top Plate)	SqFt	1.75
	Add for Stud Wall - 12" O.C.	SqFt	8%
	Deduct for Stud Wall - 24" O.C.	SqFt	22%
	Add for Bolted Plates or Sills	SqFt	.10
	Add for Each Foot Above 8'	SqFt	.06
	Add to Above for Fire Stops, Fillers and Nailers	SqFt	.42
	Add for Soffits and Suspended Framing	SqFt	.58
.13	Bridging - 1" x 3" Wood Diagonal	Set	2.40
	2" x 8" Solid	Each	2.60
.14	Rafters		
	2" x 4" Rafter (Incl Bracing) 3 - 12 Slope	SqFt	1.80
	4 - 12 Slope	SqFt	1.95
	5 - 12 Slope	SqFt	2.05
	6 - 12 Slope	SqFt	2.35
	Add for Hip-and-Valley Type	SqFt	.42
	Add for 1' 0" Overhang - Total Area of Roof	SqFt	.15
	2" x 6" Rafter (Incl Bracing) 3-12 Slope	SqFt	2.20
	4-12 Slope	SqFt	2.25
	5-12 Slope	SqFt	2.30
	6-12 Slope	SqFt	2.40
	Add for Hip - and - Valley Type	SqFt	.53
.15	Stairs	Each Floor	420.00
.16	Sub Floor Sheathing (Structural)		
	1" x 8" and 1" x 10" #3 Pine	SqFt	1.68
	1/2" x 4' x 8' CD Plywood - Exterior	SqFt	1.11
	5/8" x 4' x 8'	SqFt	1.26
	3/4" x 4' x 8'	SqFt	1.47
.17	Floor Sheathing (Over Sub Floor)		
	3/8" x 4' x 8' CD Plywood	SqFt	1.10
	1/2" x 4' x 8'	SqFt	1.18
	5/8" x 4' x 8'	SqFt	1.30
	1/2" x 4' x 8' Particle Board	SqFt	1.00
	5/8" x 4' x 8'	SqFt	1.12
	3/4" x 4' x 8'	SqFt	1.32

DIVISION #6 - WOOD & PLASTICS - QUICK ESTIMATING

0601.0	ROUGH CARPENTRY, Cont'd...	UNIT	COST	
.18	Wall Sheathing			
	1' x 8" and 1" x 10" - #3 Pine	SqFt	1.42	
	3/8" x 4' x 8' CD Plywood - Exterior	SqFt	1.08	
	1/2" x 4' x 8'	SqFt	1.17	
	5/8" x 4' x 8'	SqFt	1.37	
	3/4" x 4' x 8'	SqFt	1.60	
	25/32" x 4' x 8' Fiber Board - Impregnated	SqFt	.88	
	1" x 2' x 8' T&G Styrofoam	SqFt	1.13	
	2" x 2' x 8'	SqFt	1.66	
.19	Roof Sheathing - Flat Construction			
	1" x 6" and 1" x 8" - #3 Pine	SqFt	1.39	
	1/2" x 4' x 8' CD Plywood - Exterior	SqFt	1.20	
	5/8" x 4' x 8'	SqFt	1.32	
	3/4" x 4' x 8'	SqFt	1.50	
	Add for Sloped Roof Construction (to 5-12 slope)	SqFt	.15	
	Add for Steep Sloped Construction (over 5-12 slope)	SqFt	.27	
	Add to Above Sheathing (0601.16 thru 0601.19)			
	AC or AD Plywood	SqFt	.22	
	10' Length Plywood	SqFt	.21	
.2	HEAVY FRAMING			
.21	Columns and Beams - 16' Span Average - Floor Area	SqFt	6.80	
	20' Span	SqFt	7.05	
	24' Span	SqFt	8.15	
.22	Deck 2" x 6" T&G - Fir Random Construction Grade	SqFt	4.00	
	3" x 6" T&G - Fir Random Construction Grade	SqFt	5.25	
	4" x 6"	SqFt	6.80	
	2" x 6" T&G - Red Cedar	SqFt	5.05	
	Add for D Grade Cedar	SqFt	1.70	
	2" x 6" T&G - Panelized Fir	SqFt	3.70	
.3	MISCELLANEOUS CARPENTRY			
.31	Blocking and Bucks (2" x 4" and 2" x 6")	EACH	or	BdFt
	Doors & Windows - Nailed to Concrete or Masonry	47.15		2.30
	Bolted to Concrete or Masonry	53.30		3.20
	Roof Edges - Nailed to Wood	-		1.80
	Bolted to Concrete (Incl. bolts)	-		2.70
.32	Grounds and Furring	LnFt	or	BdFt
	1" x 4" Fastened to Wood	.75		2.25
	1" x 3"	.62		2.48
	1" x 2"	.58		3.48
	1" x 3" Fastened to Concrete/Masonry - Nailed	.95		2.85
	Gun Driven	.98		3.27
	PreClipped	1.30		3.90
	2" x 2" Suspended Framing	1.28		3.84
	Not Suspended Framing	1.05		3.15
.33	Cant Strips			
	4" x 4" Treated and Nailed	1.55		-
	Treated and Bolted (Including Bolts)	2.15		-
	6" x 6" Treated and Nailed	3.05		-
	Treated and Bolted (Including Bolts)	3.95		-
.34	Building Papers and Sealers	UNIT	COST	
	15" Felt	SqFt	.14	
	Polyethylene - 4 mil	SqFt	.13	
	6 mil	SqFt	.15	
	Sill Sealer	SqFt	.68	

DIVISION #6 - WOOD & PLASTICS - QUICK ESTIMATING

0602.0 FINISH CARPENTRY

.1   FINISH SIDINGS AND FACING MATERIALS (Exterior)

.11   Boards and Beveled Sidings

	TO WEATHER	% ADDED	SQ FT
Cedar Beveled - Clear Heart 1/2" x 4"	2 3/4"	51%	4.70
1/2" x 6"	4 3/4"	31%	4.03
1/2" x 8"	6 3/4"	23%	3.85
Rough Sawn 7/8" x 8"	6 3/4"	23%	3.60
7/8" x 10"	8 3/4"	19%	3.70
7/8" x 12"	10 3/4"	17%	3.80
3/4" Board - Rough Sawn	-	-	3.35
Redwood 1/2" x 4" Beveled - Clear Heart	2 3/4"	59%	4.05
1/2" x 6"	4 3/4"	31%	4.55
1/2" x 8"	6 3/4"	23%	4.36
5/8" x 10"	8 3/4"	19%	3.90
3/4" x 6"	4 3/4"	31%	4.70
3/4" x 8"	6 3/4"	23%	4.40
3/4" x 6" Board - Clear - 4"-6"	Varies		2.90
3/4" Tongue & Groove - 4"-6"	Varies		4.60
1" Rustic Beveled - 6" and 8"	Varies		4.83
5/4" Rustic Beveled - 6" and 8"	Varies		4.72

	UNIT	COST
Add for Metal Corners	Each	1.30
Add for Mitering Corners	Each	1.95

.12

	UNIT	COST
Plywood Fir AC Smooth -One Side 1/4"	SqFt	1.30
3/8"	SqFt	1.40
1/2"	SqFt	1.77
5/8"	SqFt	1.95
3/4"	SqFt	2.24
5/8" - Grooved and Rough Faced	SqFt	1.87
Cedar - Rough Sawn 3/8"	SqFt	1.96
5/8"	SqFt	2.22
3/4"	SqFt	2.40
5/8" - Grooved and Rough Faced	SqFt	2.45
Add for Wood Batten Strips, 1" x 2" - 4' O.C.	SqFt	.33
Add for Splines	SqFt	.35

.13   Hardboard - Paneling and Lap Siding (Primed)

		UNIT	COST
3/8" Rough Textured Paneling -	4' x 8'	SqFt	1.75
7/16" Grooved -	4' x 8'	SqFt	1.71
7/16" Stucco Board Paneling		SqFt	1.87
7/16" x 8" Lap Siding		SqFt	2.16
7/16" x 12" Lap Siding		SqFt	2.03
Add for Pre-Finishing		SqFt	.16

.14   Shingles

		UNIT	COST
Wood- 16" Red Cedar - 12" to Weather	- #1	SqFt	3.42
	- #2	SqFt	2.65
	- #3	SqFt	2.49
24" - 1/2" to 3/4" Red Cedar Hand Splits	- #1	SqFt	3.13
24" - 3/4" to 1 1/4" Red Cedar Hand Splits	- #1	SqFt	3.23
	- #2	SqFt	2.88
	- #3	SqFt	2.68
Add for Fire Retardant		SqFt	.42
Add for 3/8" Backer Board		SqFt	.70
Add for Metal Corners		Each	1.30
Add for Ridges, Hips and Corners		LnFt	3.70
Add for 8" to Weather		SqFt	25%

DIVISION #6 - WOOD & PLASTICS - QUICK ESTIMATING

			COST		
			LN FT	or	SQ FT
0602.0	**FINISH CARPENTRY, Cont'd...**				
.15	Facia - Pine #2 1" x 8"		1.60		2.40
	Cedar #3 1" x 8"		2.45		3.70
	Redwood - Clear 1" x 8"		3.50		5.25
	Plywood 5/8" - ACX 1" x 8"		1.60		2.40
.16	Soffit - Plywood 1/2" - ACX 1/2" Fir ACX				2.20
	Aluminum				2.75
	Plastic (Egg Crate)				2.90
.2	FINISH WALLS (Interior) (15% Added for Size & Cutting)				
.21	Boards, Cedar - #3 1" x 6" and 1" x 8"				3.65
	Knotty 1" x 6" and 1" x 8"				3.80
	D Grade 1" x 6" and 1" x 8"				4.53
	Aromatic 1" x 6" and 1" x 8"				4.70
	Redwood - Construction 1" x 6" and 1" x 8"				3.90
	Clear 1" x 6" and 1" x 8"				5.10
	Fir - Beaded 5/8" x 4"				3.50
	Pine - #2 1" x 6" and 1" x 8"				2.70
.22	Hardboard (Paneling) Tempered - 1/8"				1.10
	1/4"				1.20
.23	Plywood (Prefinished Paneling) 1/4" Birch - Natural				2.60
	3/4" Birch - Natural				3.45
	1/4" Birch - White				3.15
	1/4" Oak - Rotary Cut				2.50
	3/4"				3.55
	1/4" Oak - White				3.40
	1/4" Mahogany (Lauan)				1.80
	3/4"				2.70
	3/4" Mahogany (African)				4.30
	1/4" Walnut				4.10
.24	Gypsum Board (Paneling) - See 0902				

			COST	
			Prefinished Red Oak or Birch	Plastic Laminate
		UNIT		
0603.0	**MILLWORK & CUSTOM WOODWORK**			
.1	CUSTOM CABINET WORK (Red Oak or Birch)			
	Base Cabinets - Avg. 35" H x 24" D w/Drawer	LnFt	170.00	167.00
	Sink Fronts	LnFt	102.00	101.00
	Corner Cabinets	LnFt	223.00	219.00
	Add per Drawer	LnFt	43.00	43.00
	Upper Cabinets - Avg. 30" High x 12" Deep	LnFt	128.00	127.00
	Utility Cabinets - Avg. 84" High x 24" Deep	LnFt	330.00	320.00
	China or Corner Cabinets - 84" High	Each	620.00	610.00
	Oven Cabinets - 84" High x 24" Deep	Each	420.00	405.00
	Vanity Cabinets - Avg. 30" High x 21" Deep	Each	220.00	215.00
	Deduct for Prefinishing Wood	Each	12%	-
.2	COUNTER TOPS - 25"			
	Plastic Laminated with 4" Back Splash	LnFt	-	50.00
	Deduct for No Back Splash	LnFt	-	8.50
	Granite - 1 1/4" - Artificial	LnFt	-	91.00
	3/4" - Artificial	LnFt	-	71.00
	Marble	LnFt	-	97.00
	Artificial	LnFt	-	81.00
	Wood Cutting Block	LnFt	-	90.00
	Stainless Steel	LnFt	-	143.00
	Polyester-Acrylic Solid Surface	LnFt	-	160.00
	Quartz	LnFt	-	90.00
	Plastic (Polymer)	LnFt	-	60.00
	(See 1103 for Stock Cabinets)			

DIVISION #6 - WOOD & PLASTICS - QUICK ESTIMATING

0603.0 MILLWORK & CUSTOM WOODWORK, Cont'd...

.3 CUSTOM DOOR FRAMES - Including 2 Sides Trim

	UNIT	COST
Birch	Each	260.00
Fir	Each	160.00
Poplar	Each	165.00
Oak	Each	205.00
Pine	Each	195.00
Walnut	Each	370.00

See 0802.87 for Stock Prehung Units

.4 MOULDINGS AND TRIM

LINEAR FOOT

	Birch	Fir	Poplar	Oak	Pine
Apron 7/16" x 2"	2.81	-	1.82	2.40	1.90
Astragal 1 3/4" x 2 1/4"	-	-	-	5.95	4.60
Base 7/16" x 2 3/4"	3.23	1.65	1.87	2.50	2.80
9/16" x 3 1/4"	4.30	-	-	3.75	3.83
Base Shoe 7/16" x 2 3/4"	-	-	1.62	2.18	2.30
Batten Strip 5/8" x 1 5/8"	-	1.30	-	2.45	2.00
Brick Mould 1 1/14" x 2"	-	2.32	-	-	2.35
Casing 11/16" x 2 1/4"	2.98	1.80	1.98	2.82	2.55
3/4" x 3 1/2"	4.10	2.32	2.04	4.00	3.50
Chair Rail 5/8" x 1 3/4"	3.25	-	2.50	3.30	2.60
Closet Rod 1 5/16"	-	2.37	-	-	3.00
Corner Bead 1 1/8" x 1 1/8"	-	2.05	2.55	3.08	2.65
Cove Moulding 3/4" x 3/4"	2.70	-	2.18	2.60	2.20
Crown Moulding 9/16" x 3 5/8"	4.95	-	3.18	4.02	3.40
11/16" x 4 5/8"	5.75	-	-	6.30	5.35
Drip Cap	-	-	-	-	2.55
Half Round 1/2" x 1"	-	-	-	-	3.10
Hand Rail 1 5/8" x 1 3/4"	-	3.55	3.90	6.20	3.50
Hook Strip 5/8" x 2 1/2"	-	1.96	2.25	-	2.30
Picture Mould 3/4" x 1 1/2"	1.96	-	-	2.85	2.55
Quarter Round 3/4" x 3/4"	-	1.13	1.45	2.28	1.75
1/2" x 1/2"	-	1.23	-	-	1.40
Sill 3/4" x 2 1/2"	-	2.33	-	-	3.10
Stool 11/16 x 2 1/2"	5.02	2.33	3.50	5.40	3.70

	UNIT	Birch	Oak	Pine
.5 STAIRS (Treads, Risers, Skirt Boards)	LnFt	34.00	26.00	19.00
.6 SHELVING (12" Deep)	SqFt	37.00	31.00	21.00
.7 CUSTOM PANELING	SqFt	23.00	18.00	14.50
.8 THRESHOLDS - 3/4" x 3 1/2"	LnFt	-	17.00	-

0604.0 GLUE LAMINATE

	UNIT	COST
Arches 80' Span	SqFt	8.60
Beams & Purlins 40' Span	SqFt	5.90
Deck Fir - 3" x 6"	SqFt	7.40
4" x 6"	SqFt	8.05
Deck Cedar - 3" x 6"	SqFt	8.70
4" x 6"	SqFt	9.20
Deck Pine - 3" x 6"	SqFt	6.80

DIVISION #6 - WOOD & PLASTICS - QUICK ESTIMATING

0605.0 PREFABRICATED WOOD & PLYWOOD COMPONENTS

.1 WOOD TRUSSED RAFTERS -
24" O.C. - 4 -12 Pitch

		UNIT	COST		UNIT	COST
Span To 16'	w/ Supports	Each	120.00	or	SqFt	3.75
20'	and	Each	128.00		SqFt	3.20
22'	2' Overhang	Each	131.00		SqFt	2.97
24'		Each	136.00		SqFt	2.83
26'		Each	141.00		SqFt	2.72
28'		Each	156.00		SqFt	2.80
30'		Each	187.00		SqFt	3.10
32'		Each	208.00		SqFt	3.30
Add for 5-12 Pitch		Each	5%		SqFt	15%
Add for 6-12 Pitch		Each	12%		SqFt	25%
Add for Scissor Truss		Each	27.00		SqFt	.20
Add for Gable End 24'		Each	58.00		SqFt	.37

.2 WOOD FLOOR TRUSS JOISTS (TJI)
Open Web - Wood or Metal - 24" O.C.

		UNIT	COST		UNIT	COST
Up To 23' Span x 12"	Single	LnFt	7.65	or	SqFt	3.21
To 24' Span x 15"	Cord	LnFt	7.84		SqFt	3.42
To 27' Span x 18"		LnFt	8.05		SqFt	4.02
To 30' Span x 21"		LnFt	8.47		SqFt	4.24
To 25' Span x 15"	Double	LnFt	7.53		SqFt	3.86
To 30' Span x 18"	Cord	LnFt	7.84		SqFt	3.92
To 36' Span x 21"		LnFt	8.66		SqFt	4.33

Plywood Web - 24" O.C.

	UNIT	COST	UNIT	COST
Up To 15' x 9 1/2"	LnFt	4.80	SqFt	2.40
To 19' x 11 7/8"	LnFt	4.90	SqFt	2.45
To 21' x 14"	LnFt	5.50	SqFt	2.75
To 22' x 16"	LnFt	6.25	SqFt	3.13

.3 LAMINATED VENEER STRUCTURAL BEAMS

.31 Micro Lam - Plywood - 24" O.C.

	UNIT	COST		UNIT	COST
9 1/2" x 1 3/4"	LnFt	6.80	or	SqFt	3.40
11 7/8" x 1 3/4"	LnFt	7.50		SqFt	3.75
14" x 1 3/4"	LnFt	8.80		SqFt	4.40
16" x 1 3/4"	LnFt	9.80		SqFt	4.90

.32 Glue Lam- Dimension Lumber- 24" O.C.

	UNIT	COST	UNIT	COST
9" x 3 1/2"	LnFt	12.70	SqFt	6.35
12" x 3 1/2"	LnFt	14.40	SqFt	7.20
15" x 3 1/2"	LnFt	17.90	SqFt	8.85
9" x 5 1/2"	LnFt	16.60	SqFt	8.30
12" x 5 1/2"	LnFt	20.50	SqFt	10.25
15" x 5 1/2"	LnFt	25.50	SqFt	12.75
18" x 5 1/2"	LnFt	30.80	SqFt	15.40
Add for Architectural Grade	LnFt	25%		
Add for 6 3/4"	LnFt	40%		

.4 DECKS 4' x 30' w/ Plywood - Insulated — SqFt 8.80

0607.0 ROUGH HARDWARE

		UNIT	COST
.1	NAILS	SqFt or BdFt	.02
.2	JOIST HANGERS	Each	2.70
.3	BOLTS - 5/8" x 12"	Each	3.25

DIVISION #7 - THERMAL & MOISTURE PROTECTION

		PAGE
0701.0	**WATERPROOFING (07100)**	**7-2**
.1	MEMBRANE	7-2
.2	HYDROLITHIC	7-2
.3	ELASTOMERIC (Liquid)	7-2
.4	METALLIC OXIDE	7-2
.5	VINYL PLASTIC	7-2
.6	BENTONITE	7-2
0702.0	**DAMPPROOFING (07150)**	**7-2**
.1	BITUMINOUS	7-2
.2	CEMENTITIOUS	7-2
.3	SILICONE	7-2
0703.0	**BUILDING INSULATION (07210)**	**7-2**
.1	FLEXIBLE (Fibrous Batts & Rolls)	7-2
.2	RIGID	7-3
.3	LOOSE	7-3
.4	FOAMED	7-3
.5	SPRAYED	7-3
.6	ALUMINUM	7-3
.7	VINYL FACED FIBERGLASS	7-3
0704.0	**ROOF & DECK INSULATION (07240)**	**7-4**
0705.0	**MEMBRANE ROOFING (07500)**	**7-4**
.1	BUILT UP	7-4
0706.0	**SINGLE PLY ROOFING**	**7-4**
0707.0	**TRAFFIC ROOF COATING (07570)**	**7-5**
.1	PEDESTRIAN	7-5
.2	VEHICULAR	7-5
0708.0	**SHINGLE & TILE ROOFING (07300)**	**7-6**
.1	ASPHALT SHINGLES	7-6
.2	FIBERGLASS SHINGLES	7-6
.3	WOOD SHINGLES	7-6
.4	METAL SHINGLES	7-6
.5	SLATE ROOFING	7-6
.6	TILE ROOFING	7-6
.61	Clay	7-6
.62	Concrete	7-6
.63	Steel	7-6
.7	CORRUGATED ROOFING	7-6
0709.0	**SHEET METAL & FLASHING (07600)**	**7-6**
.1	DOWNSPOUTS, GUTTERS & FITTINGS	7-6
.2	FACINGS, GRAVEL STOPS & TRIM	7-7
.3	FLASHINGS	7-7
.4	SHEET METAL ROOFING	7-7
.5	EXPANSION JOINTS	7-7
0710.0	**ROOF ACCESSORIES (07700)**	**7-8**
.1	SKYLIGHTS	7-8
.2	SKY DOMES	7-8
.3	GRAVITY VENTILATORS	7-8
.4	SMOKE VENTILATORS	7-8
.5	PREFAB CURBS & EXPANSION JOINTS	7-8
.6	ROOF LINE LOUVRES	7-8
0711.0	**SEALANTS (07900)**	**7-8**
.1	CAULKING	7-8
.2	JOINT FILLERS & GASKETS	7-8
7A	**QUICK ESTIMATING**	**7A-9 & 7A-10**

DIVISION #7 - THERMAL & MOISTURE PROTECTION

		UNIT	LABOR	MATERIAL
0701.0	**WATERPROOFING (L&M) (Roofers)**			
.1	MEMBRANE			
	1 Ply Membrane - Felt 15#	SqFt	.47	.34
	Fabric and Felt	SqFt	.67	.38
	2 Ply Membrane - Felt 15#	SqFt	.81	.55
	Fabric and Felt	SqFt	.94	.59
.2	HYDROLITHIC	SqFt	.82	.83
.3	ELASTOMERIC- Rubberized Asphalt .060M & Poly	SqFt	.82	.56
	Liquid	SqFt	.83	.69
.4	METALLIC OXIDE- 3 Coat	SqFt	.77	1.76
	3 Coat with Sand & Cement Cover	SqFt	.99	1.94
.5	VINYL PLASTIC	SqFt	.50	1.05
.6	BENTONITE - 3/8"	SqFt	.67	.62
0702.0	**DAMPPROOFING (Roofers or Laborers)**			
.1	BITUMINOUS (Asphalt)			
.11	Trowel Mastics			
	Asphalt 1/16" (12 Sq.Ft. Gal.)	SqFt	.29	.22
	1/8" (20 Sq.Ft. Gal.)	SqFt	.34	.27
.12	Spray or Brush Liquid			
	Asphalt - 1 Coat - Spray	SqFt	.20	.14
	Spray	SqFt	.26	.19
	Brush	SqFt	.30	.21
.13	Hot Mop			
	Asphalt - 1 Coat - Including Primer	SqFt	.48	.36
	Fibrous Asphalt - 1 Coat	SqFt	.54	.39
.14	Trowel Mastics & Pre-formed Vapor Barriers			
	Mastic Fabric Mastic	SqFt	.50	.75
	and Polyvinyl Chloride	SqFt	.48	.62
.2	CEMENTITIOUS - 1 Coat - 1/2"	SqFt	.29	.15
	2 Coat - 1"	SqFt	.40	.19
.3	SILICONE (Masonry) - 1 Coat	SqFt	.22	.16
	2 Coat	SqFt	.29	.26
	Add for Scaffold to Above Sections	SqFt	.21	.11
0703.0	**BUILDING INSULATION (S) (Carpenters)**			
.1	FLEXIBLE (Batts and Rolls) - Friction			
	Fiberglass			
	2 1/4" R 7.4	SqFt	.23	.24
	3 1/2" R 11.0	SqFt	.24	.26
	3 5/8" R 13	SqFt.	.25	.33
	6" R 19.0	SqFt	.27	.39
	9" R 30.0	SqFt	.31	.66
	Add for Paper Faced Batts	SqFt	.02	.03
	Add for Polyethylene Barrier 2 Mil	SqFt	.04	.04
	Add for Staggered Stud Installation	SqFt	.08	.03

DIVISION #7 - THERMAL & MOISTURE PROTECTION

0703.0 BUILDING INSULATION, Cont'd...

		UNIT	LABOR	MATERIAL
.2	**RIGID**			
	Fiberglass			
	1" R 4.35 3# Density	SqFt	.28	.31
	1 1/2" R 6.52 3# Density	SqFt	.30	.49
	2" R 8.70 3# Density	SqFt	.32	.64
	Add for 4.2# Density per Inch	SqFt	-	.07
	Add for 6.0# Density per Inch	SqFt	-	.25
	Add for Kraft Paper Backed	SqFt	-	.08
	Add for Aluminum Foil Backed	SqFt	-	.35
	Add for Fireproof Type	SqFt	-	10%
	Expanded Styrene			
	Molded, White (Bead Board)			
	1" x 16" x 8' R 5.40	SqFt	.28	.20
	1" x 24" x 8', T&G R 5.40	SqFt	.31	.27
	1 1/2" R 6.52	SqFt	.32	.29
	2" R 7.69	SqFt	.35	.37
	2" x 24" x 8', T&G R 7.69	SqFt	.35	.53
	Extruded, Blue			
	1" x 24" x 8' R 5.40	SqFt	.28	.39
	T&G	SqFt	.31	.49
	1 1/2" R 6.52	SqFt	.30	.49
	2" R 7.69	SqFt	.32	.51
	2" x 24" x 8', T&G R 7.69	SqFt	.35	.76
	Perlite			
	1" R 2.78	SqFt	.28	.44
	2" R 5.56	SqFt	.34	.77
	Urethane			
	1" R 6.67	SqFt	.31	.80
	2" R 13.34	SqFt	.36	1.48
.3	**LOOSE**			
	Expanded Styrene 1" R 3.80	SqFt	.22	.22
	Fiberglass 1" R 2.20	SqFt	.23	.23
	Rockwool 1" R 2.90	SqFt	.24	.23
	Cellulose 1" R 3.70	SqFt	.22	.24
.4	**FOAMED**			
	Urethane per Inch or Bd.Ft.	SqFt	.45	.82
.5	**SPRAYED**			
	Fibered Cellulose per Inch or Bd.Ft.	SqFt	.40	.43
	Polystyrene per Inch or Bd.Ft.	SqFt	.47	.51
	Urethane per Inch or Bd.Ft.	SqFt	.52	.92
.6	**ALUMINUM PAPER**	SqFt	.15	.14
.7	**VINYL FACED FIBERGLASS**			
	2 5/8" x 63 1/2" x 96" R 10.0	SqFt	.36	.87
	Add to All Above for Ceiling Work	SqFt	.06	-
	Add to All Above for Clip On	SqFt	.07	.02
	Add to All Above for Scaffold	SqFt	.19	.10

See Division 3 for Cementitious Poured and Rigid Insulation
See Division 4 for Core and Cavity Filled Masonry
See Division 7 for Roof Insulation
See Division 15 for Mechanical Insulations

DIVISION #7 - THERMAL & MOISTURE PROTECTION

		UNIT	COST
0704.0	**ROOF AND DECK INSULATION (L&M) (Roofers)**		
.1	RIGID - NON RATED		
	Fiber Board 1/2" R 1.39	Sq	46.90
	3/4" R 2.09	Sq	57.10
	1" R 2.78	Sq	66.30
	Fiberglass 1" R 3.70	Sq	76.50
	1 1/2" R 2.09	Sq	96.90
	Perlite 1/2" R 1.39	Sq	45.90
	3/4" R 2.09	Sq	57.10
	1" R 2.78	Sq	67.30
	2" R 5.26	Sq	100.00
	Expanded Polystrene 1" R 4.35	Sq	61.20
	1 1/2" R 6.52	Sq	71.40
	2" R 7.69	Sq	89.80
.2	RIGID - FIRE RATED		
	Isocyanurate 1" R 6.30 Glass Faced	Sq	102.00
	1 1/5" R 7.60	Sq	107.10
	1 1/2" R 10.00	Sq	112.20
	1 3/4" R 12.00	Sq	122.40
	2" R 14.00	Sq	127.50
	Phenolic Foam 1" R 7.60	Sq	84.60
	1 1/2" R 12.50	Sq	94.90
	2" R 16.60	Sq	117.00
	Add for 5/8" Sheet Rock over Steel Deck	Sq	13.50
	Add for Fiber or Mineral Cants	LnFt	1.45
	Add for Felt Faces	Sq	9.50
	Add for Composite Insulations	Sq	25%
.3	SPRAYED - 2" Polystyrene	Sq	230.00
0705.0	**MEMBRANE ROOFING (L&M) (Roofers)**		
	BUILT-UP ROOFING (50 SQUARES OR MORE)		
	Asphalt & Gravel - 15# Glass/Organic Felts 3-Ply	Sq	140.00
	4 - Ply	Sq	150.00
	5 - Ply	Sq	170.00
	Add for Pitch and Gravel	Sq	42.00
	Add for Curbed Openings	Each	121.00
	Add for Round Vent Openings	Each	110.00
	Add for Roof Skids	LnFt	24.20
	Add for Sloped Roofs	Sq	31.50
	Add for Applications on Wood	Sq	10.50
	Add for Marble Chips	Sq	47.00
	Add for Barriers - 15" Felt	Sq	9.20
	Fire Resistant	Sq	21.00
	Add for Walkways - 1"	SqFt	4.46
	Add for Bond	Sq	10.50
	Add for Roof Openings Cut in Existing	Each	400.00

See Quick Estimating 7A-10 for Combined Roofing, Insulation and Sheet Metal

DIVISION #7 - THERMAL & MOISTURE PROTECTION

0706.0 SINGLE-PLY ROOFING No Insulation or Flashing Included

	UNIT	Gravel Ballast	Application Mechanical Fastened	Adhesive Fastened
Butylene 100 mil	Sq	172.00	187.00	192.00
EPDM 50 mil	Sq	135.00	161.00	187.00
Neoprene 60 mil	Sq	250.00	300.00	210.00
P.V.C. 50 mil	Sq	125.00	182.00	200.00
Add per Opening	Opng	120.00	-	-

0707.0 TRAFFIC ROOF COATINGS (L&M)

	UNIT	LABOR	MATERIAL
.1 PEDESTRIAN TRAFFIC TYPE	SqFt	-	6.00
.2 VEHICULAR TRAFFIC TYPE			
Rubberized Coating with 3/4" Asphalt Topping	SqFt	-	4.04
Polyurethane with Non-Slip Aggregates on 20 mil silicone rubber	SqFt	-	3.62
Elastomeric with Non-Skid - 2 Coat	SqFt	-	2.31
3 Coat	SqFt	-	3.05

0708.0 SHINGLE, TILE and CORRUGATED ROOFING

	UNIT	LABOR	MATERIAL
.1 ASPHALT SHINGLES (S) (Carpenters) 4-12 Pitch			
Square Butt 235#- Seal Down	SqFt	.40	.35
300# - Laminated	SqFt	.45	.59
325# - Fire Resistant	SqFt	.48	.57
Timberline 300# - 25 Year	SqFt	.54	.48
300# - 30 Year	SqFt	.56	.66
Tab Lock 340#	SqFt	.54	.88
Roll 90# - 100 SqFt per Roll	SqFt	.25	.18
55# - 100 SqFt per Roll	SqFt	.21	.15
Base Starter - 240 SqFt per Roll	SqFt	.21	.22
Ice & Water Starter - 100 SqFt per Roll	SqFt	.26	.40
Add for Boston Ridge Shingle	LnFt	1.04	.44
Add for 15# Felt Underlayment	SqFt	.06	.04
Add for Pitches 6-12 & Up - Ea Pitch Increase	SqFt	.06	-
Add for Removal and Haul - Away (2 Layers)	SqFt	.28	.12
Add for Chimneys, Skylights & Bay Windows	Each	41.60	-
.2 FIBERGLASS SHINGLES - 215# - Seal Down	SqFt	.40	.34
250# - Seal Down	SqFt	.45	.52
.3 WOOD			
Sawed			
Red Cedar - 16" x 5" to Weather - #1 Grade	SqFt	.73	1.50
#2 Grade	SqFt	.71	1.20
#3 Grade	SqFt	.67	.75
Hand Splits			
Red Cedar - 1/2" x 3/4", 24" x 10" to Weather	SqFt	.87	1.20
3/4"x 1 1/4", 24" x 10" to Weather	SqFt	.94	1.50
Fancy Butt - 7" to Weather	SqFt	1.04	3.20
Add for Fire Retardant	SqFt	-	.40
Add for Pitches over 5-12, Each Pitch Increase	SqFt	.06	-
.4 METAL			
Aluminum - Mill .020	SqFt	.77	1.50
.030	SqFt	.78	1.65
Anodized .020	SqFt	.77	1.90
.030	SqFt	.78	2.15
Steel - Enameled - Colored	SqFt	.82	2.60
Galvanized - Colored	SqFt	.82	2.00

DIVISION #7 - THERMAL & MOISTURE PROTECTION

0708.0 SHINGLE, TILE AND CORRUGATED ROOFING, Cont'd...

		UNIT	COST
.6	SLATE SHINGLES (Incl. Underlayment) (L&M) (Roofers)		
	Vermont and Pennsylvania		
	3/16" - 6 1/2" to 8 1/2" to Weather		
	Weathered Green or Black	Sq	1,000.00
	Gray	Sq	1,000.00
	Fading Purple	Sq	1,000.00
	Unfading Purple	Sq	1,300.00
	Add for Variegated Colors & Sizes - 1/4" to 1/2"	Sq	340.00
	3/8" to 1"	Sq	650.00
	Add for Hips and Valleys	LnFt	51.00
	Add for Pitches over 6-12	Sq	71.00
.7	TILE (L&M) (Roofers)		
.71	Clay, Interlocking Shingle Type - Designed	Sq	990.00
	Early American	Sq	1,000.00
	Williamsburg	Sq	1,100.00
	Mormon	Sq	1,600.00
	Architectural Pattern Shingle Type - Provincial	Sq	950.00
	Georgian	Sq	950.00
	Colonial	Sq	950.00
	Spanish, Red	Sq	680.00
	Mission	Sq	1,100.00
.72	Concrete - Plain - 10" to Weather - 9" Wide	Sq	600.00
	Colored	Sq	670.00
	Add for Staggered Installation	Sq	220.00
.73	Steel - 16 Ga - Ceramic Coated	Sq	410.00
.74	Aluminum .019	Sq	480.00
.8	CORRUGATED PANELS		
.81	Concrete - Plain	Sq	3.47
.82	Vinyl - 120	Sq	5.50
.83	Fiberglass - 8 oz	Sq	3.67
.84	Aluminum .024 - Plain	Sq	3.11
	.024 - Painted	Sq	3.77
	See Division 5 for Corrugated Metal Panels		

0709.0 SHEET METAL WORK (L&M)

(Sheet Metal Workers)

			COST			
		UNIT	Galvanized 25 Ga	Aluminum .032"	Copper 16 oz	Enameled
.1	DOWNSPOUTS, GUTTERS AND FITTINGS					
	Downspouts - 3" x 2"	LnFt	2.35	2.35	6.40	2.90
	4" x 3"	LnFt	2.90	3.10	9.10	3.90
	5" x 4"	LnFt	4.00	4.10	9.50	5.00
	Round 3"	LnFt	2.25	2.75	5.90	2.90
	4"	LnFt	2.65	3.30	7.00	3.70
	5"	LnFt	3.35	4.00	7.80	4.45
	Chute Type 4" x 6"	LnFt	3.85	4.50	-	5.40
	5" x 7"	LnFt	4.40	5.05	-	6.20
	Gutters - Box or Half Round 4"	LnFt	3.10	4.00	6.40	4.20
	5"	LnFt	3.60	4.10	7.90	4.65
	6"	LnFt	4.40	4.70	8.10	5.70
	Add for Guards	LnFt	-	1.00	-	-
	Fittings (Elbows, Corners, etc.)	LnFt	9.30	8.80	11.00	11.00

See Quick Estimating 7A-10 for Combined Roofing, Insulation and Sheet Metal.

DIVISION #7 - THERMAL & MOISTURE PROTECTION

0709.0 SHEET METAL WORK, Cont'd...

.2 FASCIAS, GRAVEL STOPS & TRIM

	UNIT	Galvanized 25 Ga	Aluminum .032"	Copper 16 oz	Enameled	PVC
Fascias - 6"	LnFt	4.40	6.80	9.80	5.40	5.90
8"	LnFt	4.70	7.60	10.80	5.70	6.50
Copings - 12"	LnFt	5.60	9.00	11.30	6.80	-
Scuppers	Each	105.00	105.00	125.00	110.00	-
Gravel Stops - 4"	LnFt	4.30	6.30	7.40	5.50	-
6"	LnFt	4.00	7.10	9.30	5.70	6.10
8"	LnFt	5.20	7.50	10.50	6.30	6.50
Add for Durodonic Finish	LnFt		25%			

.3 FLASHINGS

	UNIT	COST
Galvanized - 25 ga	SqFt	4.30
Aluminum - .019	SqFt	4.00
.032	SqFt	4.30
Copper - 16 oz	SqFt	7.00
20 oz	SqFt	7.20
Asphalted Fabric - Plain	SqFt	2.15
Copper Backed	SqFt	3.50
Aluminum Back	SqFt	3.10
Vinyl Fabric - .020	SqFt	1.15
.030	SqFt	1.30
Lead - 2.5#	SqFt	6.50
Rubber - 1/16"	SqFt	2.60
PVC - 24 mil	SqFt	2.90

.4 SHEET METAL ROOFING - Including Underlayment

	UNIT	COST
Copper - Standing Seam - 16 oz	Sq	910.00
20 oz	Sq	1,030.00
Batten Seam 16 oz	Sq	920.00
20 oz	Sq	1,050.00
Flat Lock 16 oz	Sq	850.00
20 oz	Sq	970.00
Stainless Steel - Batten Seam 28 ga	Sq	930.00
Standing Seam 28 ga	Sq	1,030.00
Flat Seam 28 ga	Sq	880.00
Monel - Batten Seam .018"	Sq	1,050.00
Standing Seam .018"	Sq	1,080.00
Flat Seam .108"	Sq	1,000.00
Lead - Copper Coated - Batten Seam 3 lb	Sq	990.00
Flat Seam 3 lb	Sq	930.00
Aluminum - Clad	Sq	760.00
Galvanized - Colored .020"	Sq	830.00

See Division 5 for Corrugated Type

.5 EXPANSION JOINTS - 2 1/2"

	UNIT	COST
Polyethylene with galvanized	LnFt	6.30
Galvanized 25 ga	LnFt	7.40
Aluminum .032	LnFt	8.40
Copper 16 oz	LnFt	20.40
Stainless Steel	LnFt	11.10
Neoprene	LnFt	8.10
Butyl	LnFt	6.50
Add for 3 1/2"	LnFt	10%

DIVISION #7 - THERMAL & MOISTURE PROTECTION

0710.0	ROOF AND SOFFITT ACCESSORIES (L&M) (Roofers)	UNIT	COST	or	UNIT	COST
.1	SKYLIGHTS					
	Example: 4' - 0" x 4' - 0"	SqFt	35.70		Each	570.00
	8' - 0" x 8' - 0"	SqFt	22.40		Each	1,400.00
.2	SKY DOMES (Plastic - Clear or White)					
	20" x 20" Single				Each	158.00
	Double				Each	189.00
	24" x 24" Single				Each	194.00
	Double				Each	220.00
	48" x 48" Single				Each	460.00
	Double				Each	530.00
	72" x 72" Single				Each	780.00
	Double				Each	950.00
	96" x 96" Double				Each	1,500.00
	Add for Bronze Top Colors				-	10%
	Add for Electric Operated Type				Each	440.00
	Add for Pyramid Type				Each	100%
	Add for Hip Type				Each	50%
.3	GRAVITY VENTILATORS					
	Diameter Aluminum or Galvanized 12"				Each	100.00
	24"				Each	240.00
.4	SMOKE VENTILATORS					
	Example: 48" x 96" - Aluminum				Each	2,040.00
	48" x 48" - Steel				Each	1,900.00
.5	PREFAB CURBS & EXPANSION JOINTS - 3" Fibered				LnFt	1.58
	4" Fibered				LnFt	1.63
.6	ROOF LINE LOUVRES				LnFt	1.53
.7	TURBINE VENTILATORS - 12"				Each	310.00
	24"				Each	320.00
.8	CUPOLAS				Each	230.00
.9	ROOF VENTS - P.V.C.				Each	58.00
	Aluminum				Each	53.00
	See 1012 for Roof Hatches					
0711.0	SEALANTS (L&M) (Carpenters)					
.1	CAULKING - 1/2" x 1/2" Joints					
	Windows and Doors - 2-Part Proprietary (Oil Base)				LnFt	1.70
	Silicone Rubber				LnFt	1.96
	Polysulfide (Thiakol)				LnFt	2.11
	Acrylic				LnFt	2.06
	Polyurethane				LnFt	2.06
	Control Joints - 2 - Part Proprietary				LnFt	1.85
	Silicone Rubber				LnFt	2.21
	Polysulfide				LnFt	3.04
	Acrylic				LnFt	2.21
	Polyurethane				LnFt	2.21
	Stone Pointing - Silicone Rubber				LnFt	2.52
	Polysulfide				LnFt	2.78
	Acrylic				LnFt	2.21
	Add for Swing Stage Work				LnFt	.41
	Add for 3/4" x 3/4" Joints				LnFt	.82
	Deduct for 1/4" x 1/4" Joints				LnFt	.26
.2	GASKETS AND JOINT FILLERS					
	Gaskets - 1/4" Polyvinyl x 6"				LnFt	2.06
	1/2" Polyvinyl x 6"				LnFt	2.52
	1/4" Neoprene x 6"				LnFt	2.68
	1/2" Neoprene x 6"				LnFt	3.50
	Joint Fillers -1/2"				LnFt	.31
	3/4"				LnFt	.35
	1"				LnFt	.39

DIVISION #7 - THERMAL & MOISTURE PROTECTION - QUICK ESTIMATING

		SQ.FT. COST
0701.0	**WATERPROOFING**	
	1-Ply Membrane Felt 15#	1.30
	2-Ply Membrane Felt 15#	2.00
	Hydrolithic	2.24
	Elastomeric Rubberized Asphalt with Poly Sheet	2.03
	Liquid	2.24
	Metallic Oxide 3-Coat	3.33
	Vinyl Plastic	2.29
	Bentonite 3/8" - Trowel	1.61
	5/8" - Panels	1.82
0702.0	**DAMPPROOFING**	
	Asphalt Trowel Mastic 1/16"	.78
	1/8"	.93
	Spray Liquid 1-Coat	.49
	2-Coat	.68
	Brush Liquid 2-Coat	.76
	Hot Mop 1-Coat and Primer	1.30
	1 Fibrous Asphalt	1.40
	Cementitious Per Coat 1/2" Coat	.64
	Silicone 1-Coat	.47
	2-Coat	.71
	Add for Scaffold and Lift Operations	.42
0703.0	**BUILDING INSULATION**	
	FLEXIBLE	
	Fiberglass 2 1/4" R 7.40	.66
	3 1/2" R 11.00	.70
	3 5/8" R 13.00	.78
	6" R 19.00	.90
	9" R 30.00	1.24
	12" R 38.00	1.49
	Rock Wool 3 1/2" R 11.00	.66
	4" R 13.00	.73
	6" to 7" R 19.00	.90
	9" R 30.00	1.24
	Add for Ceiling Work	.08
	Add for Paper Faced	.08
	Add for Polystyrene Barrier (2m)	.08
	Add for Scaffold Work	.42
	RIGID	
	Fiberglass 1" R 4.35 3# Density	.83
	1 1/2" R 6.52 3# Density	1.09
	2" R 8.70 3# Density	1.34
	Styrene, Molded 1" R 4.35	.66
	1 1/2" R 6.52	.76
	2" R 7.69	.91
	2" T&G R 7.69	1.01
	Styrene, Extruded 1" R 5.40	.96
	1 1/2" R 6.52	1.08
	2" R 7.69	1.34
	2" T&G R 7.69	1.50
	Perlite 1" R 2.78	.98
	2" R 5.56	1.50
	Urethane 1" R 6.67	1.18
	2" R 13.34	2.21
	Add for Glued Applications	.08

DIVISION #7 - THERMAL & MOISTURE PROTECTION - QUICK ESTIMATING

		SQ.FT.COST
0703.0	**BUILDING INSULATION, Cont'd...**	
	LOOSE 1" Styrene R 3.8	.60
	1" Fiberglass R 2.2	.56
	1" Rock Wool R 2.9	.56
	1" Cellulose R 3.7	.52
	FOAMED Urethane Per Inch	1.55
	SPRAYED Cellulose Per Inch	1.08
	Polystyrene	1.29
	Urethane	1.96
	ALUMINUM PAPER	1.60
	VINYL FACED FIBERGLASS	1.75
0704.0	**ROOF & DECK INSULATION** - COMBINED WITH 0709.0 BELOW	
0705.0	**MEMBRANE ROOFING** - COMBINED WITH 0709.0 BELOW	
0708.0	**SHINGLE ROOFING**	
	ASPHALT SHINGLES 235# Seal Down	1.07
	300# Laminated	1.39
	325" Fire Resistant	1.50
	Timberline	1.71
	340# Tab Lock	1.82
	ASPHALT ROLL 90#	.54
	55#	.43
	FIBERGLASS SHINGLES 215#	1.05
	250#	1.35
	RED CEDAR 16" x 5" to Weather #1 Grade	2.88
	#2 Grade	2.27
	#3 Grade	1.96
	24" x 10" to Weather - Hand Splits - 1/2" x 3/4" #2	2.58
	3/4" x 1 1/4" #2	2.99
	Add for Fire Retardant	.52
	METAL Aluminum 020 Mill	2.73
	020 Anodized	3.30
	Steel, Enameled Colored	4.12
	Galvanized Colored	3.35
	Add to Above for 15# Felt Underlayment	
	Add for Pitches Over 5-12 each Pitch Increase	.06
	Add for Chimneys, Skylights and Bay Windows - Each	60.00
0704.0/ 0705.0/ 0709.0 - COMBINED ROOFING INSULATION and SHEET METAL		
		SQ. COST
	MEMBRANE	
	3-Ply Asphalt & Gravel - Fiberboard & Perlite - R 10.0	400.00
	R 16.6	450.00
	4-Ply - R 10.0	460.00
	R 16.6	480.00
	5-Ply - R 10.0	520.00
	R 16.6	570.00
	Add for Sheet Rock over Steel Deck - 5/8"	108.00
	Add for Fiberglass Insulation	34.60
	Add for Pitch and Gravel	54.00
	Add for Sloped Roofs	43.20
	Add for Thermal Barrier	34.60
	Add for Upside Down Roofing System	99.50
	SINGLE-PLY - 60M Butylene Roofing & Gravel Ballast - R 10.0	370.00
	Mech. Fastened - R 10.0	390.00
	PVC & EPDM Roofing & Gravel Ballast - R 10.0	341.00
	Mech. Fastened - R 10.0	380.00
	Blocking and Cants Not Included	

DIVISION #8 - DOORS, WINDOWS & GLASS

			PAGE
0801.0	**HOLLOW METAL DOORS & FRAMES (CSI 08100)**		**8-3**
.1	CUSTOM		8-3
.11		Frames	8-3
.12		Doors	8-3
.2	STOCK		8-3
.21		Frames	8-3
.22		Doors	8-3
0802.0	**WOOD DOORS (CSI 08200)**		**8-4**
.1	FLUSH DOORS		8-4
.2	PANEL DOORS		8-4
.3	LOUVERED DOORS		8-5
.4	BI-FOLD DOORS - FLUSH		8-5
.5	CAFE DOORS		8-5
.6	FRENCH DOORS		8-5
.7	DUTCH DOORS		8-5
.8	PREHUNG DOOR UNITS		8-5
0803.0	**SPECIAL DOORS (CSI 08300)**		**8-6**
.1	BLAST OR NUCLEAR RESISTANT (CSI 08315)		8-6
.2	CLOSET - BI-FOLDING (CSI 08350)		8-6
.21		Wood	8-6
.22		Metal	8-6
.23		Leaded Metal	8-6
.3	FLEXIBLE DOORS (CSI 08350)		8-6
.4	GLASS - ALL (CSI 08340)		8-6
.5	HANGAR & INDUSTRIAL (CSI 08370)		8-6
.6	METAL COVERED FIRE & INDUSTRIAL SLIDING DOORS (CSI 08300)		8-6
.7	OVERHEAD DOORS - COMMERCIAL & RESIDENTIAL (CSI 08360)		8-7
.8	PLASTIC LAMINATE FACED (CSI 08220)		8-7
.9	REVOLVING DOORS (CSI 08470)		8-7
.10	ROLLING DOORS & GRILLES (CSI 08330 & 08350)		8-8
.11	SHOWER DOORS (CSI 10820)		8-8
.12	SLIDING OR PATIO DOORS (CSI 08640)		8-8
.13	SOUND REDUCTION (CSI 08385)		8-8
.14	TRAFFIC (CSI 08380)		8-8
.15	SPECIAL MADE & ENGINEERED INDUSTRIAL DOORS (CSI 08370)		8-8
.16	VAULT DOORS (CSI 11030)		8-8
0804.0	**ENTRANCE DOORS, FRAMES & STORE FRONT CONSTRUCTION (CSI 08400)**		**8-9**
.1	ALUMINUM		8-9
.11		Stock	8-9
.12		Custom	8-9
.2	BRONZE		8-9
.3	STAINLESS STEEL		8-9
0805.0	**METAL WINDOWS (CSI 08500)**		**8-9**
.1	ALUMINUM WINDOWS		8-9
.2	ALUMINUM SASH		8-9
.3	STEEL WINDOWS		8-9
.4	STEEL SASH		8-9
.5	AWNING		8-9
0806.0	**WOOD & VINYL CLAD WINDOWS (CSI 08610)**		**8-10**
.1	BASEMENT OR UTILITY		8-10
.2	CASEMENT OR AWNING		8-10
.3	DOUBLE HUNG		8-10
.4	GLIDER WITH SCREEN		8-10
.5	PICTURE WINDOW WITH SCREENS		8-11
.6	CASEMENT ANGLE BAY WINDOWS		8-11
.7	CASEMENT BOW WINDOWS		8-11
.8	90° CASEMENT BOX BAY WINDOWS		8-11
.9	SKY OR ROOF WINDOWS		8-11
.10	CIRCLE TOPS AND ROUND		8-11

DIVISION #8 - DOORS, WINDOWS & GLASS

		PAGE
0807.0	**SPECIAL WINDOWS (CSI 08650)**	**8-12**
.1	LIGHT-PROOF WINDOWS (CSI 08667)	8-12
.2	PASS WINDOWS (CSI 08665)	8-12
.3	DETENTION WINDOWS (CSI 08660)	8-12
.4	VENETIAN BLIND WINDOWS	8-12
.5	SOUND CONTROL WINDOWS (CSI 08653)	8-12
0808.0	**DOOR AND WINDOW ACCESSORIES**	**8-12**
.1	STORMS AND SCREENS (CSI 08670)	8-12
.2	DETENTION SCREENS (CSI 08660)	8-12
.3	DOOR OPENING ASSEMBLIES (CSI 08720)	8-12
.4	SHUTTERS - FOLDING (CSI 08668) & ROLL-UP (CSI 08664)	8-12
0809.0	**CURTAIN WALL SYSTEMS (CSI 08900)**	**8-12**
.1	STRUCTURAL OR VERTICAL TYPE	8-12
.2	PANEL WALL OR HORIZONTAL TYPE	8-12
.3	ARCHITECTURAL PANELS	8-12
0810.0	**FINISH HARDWARE (CSI 08710)**	**8-13**
.1	BUTTS	8-13
.2	CATCHES - ROLLER	8-13
.3	CLOSURES	8-13
.4	DEAD BOLT LOCKS	8-13
.5	EXIT DEVICES	8-13
.6	HINGES, SPRING	8-13
.7	LATCHSETS	8-13
.8	LOCKSETS	8-13
.9	PLATES	8-13
.10	STOPS AND HOLDERS	8-13
0811.0	**WEATHER-STRIPPING & THRESHOLDS (CSI 08730)**	**8-13**
.1	ASTRAGALS	8-13
.2	DOORS	8-13
.3	SWEEPS	8-13
.4	THRESHOLDS	8-13
.5	WINDOWS	8-13
0812.0	**GLASS AND GLAZING (CSI 08810)**	**8-14**
.1	BEVELED GLASS	8-14
.2	COATED	8-14
.3	HEAT ABSORBING, et al	8-14
.4	INSULATED GLASS	8-14
.5	LAMINATED (SAFETY AND SOUND)	8-14
.6	MIRROR GLASS	8-14
.7	PATTERNED OR ROUGH (OBSCURE)	8-14
.8	PLASTIC GLASS (SAFETY)	8-14
.9	PLATE - CLEAR	8-14
.10	SANDBLASTED	8-14
.11	SHEET GLASS	8-14
.12	STAINED GLASS	8-14
.13	STRUCTURAL GLASS	8-14
.14	TEMPERED (SAFETY PLATE)	8-14
.15	TINTED PLATE - LIGHT REFLECTIVE	8-14
.16	WIRED (SAFETY)	8-14
8A	**QUICK ESTIMATING SECTIONS**	**8A-15 thru 8A-20**

DIVISION #8 - DOORS, WINDOWS & GLASS

0801.0 HOLLOW METAL (M) (CARPENTERS)

		UNIT	LABOR	MATERIAL
.1	CUSTOM			
.11	Frames (16 Gauge)			
	2'6" x 6'8" x 4 3/4"	Each	46	134
	2'8" x 6'8" x 4 3/4"	Each	47	144
	3'0" x 6'8" x 4 3/4"	Each	48	155
	3'4" x 6'8" x 4 3/4"	Each	57	160
	3'6" x 6'8" x 4 3/4"	Each	64	165
	4'0" x 6'8" x 4 3/4"	Each	73	175
	Deduct for 18 Gauge	Each	8	15
	Add for 14 Gauge	Each	11	23
	Add for A, B, or C Label	Each	7	12
	Add for Transoms	Each	28	72
	Add for Side Lights	Each	30	88
	Add for Frames over 7'0" (Height)	Each	30	15
	Add for Frames over 6 3/4" (Width)	Each	12	15
	Add for Bolted Frame with Bolts	Each	21	31
.12	Doors (1 3/4" - 18 Gauge)			
	2'6" x 6'8" x 1 3/4"	Each	49	165
	2'8" x 6'8" x 1 3/4"	Each	50	175
	3'0" x 6'8" x 1 3/4"	Each	52	185
	3'4" x 6'8" x 1 3/4"	Each	59	196
	3'6" x 6'8" x 1 3/4"	Each	67	210
	4'0" x 6'8" x 1 3/4"	Each	78	230
	Add for 16 Gauge	Each	11	36
	Deduct for 20 Gauge	Each	7	15
	Add for A Label	Each	8	15
	Add for B or C Label	Each	7	15
	Add for Vision Panels or Lights	Each	-	67
	Add for Louvre Openings	Each	-	62
	Add for Galvanizing	Each	-	41
.2	STOCK			
.21	Frames (4 3/4"-5 3/4"-6 3/4"-8 3/4"- 16 Ga)			
	2'6" x 6'8" or 7'0"	Each	45	82
	2'8" x 6'8" or 7'0"	Each	46	88
	3'0" x 6'8" or 7'0"	Each	49	93
	3'4" x 6'8" or 7'0"	Each	55	98
	3'6" x 6'8" or 7'0"	Each	62	103
	4'0" x 6'8" or 7'0"	Each	71	124
	See above .11 for Size Change			
	Add for Bolted Frame & Bolts	Each	21	31
	Add for A, B, or C Label	Each	8	15
	Deduct Non-Welded Frame (Knock Down)	Each	5	15
.22	Doors (1 3/8" or 1 3/4" - 18 Gauge)			
	2'6" x 6'8" or 7'0"	Each	49	170
	2'8" x 6'8" or 7'0"	Each	50	175
	3'0" x 6'8" or 7'0"	Each	52	185
	3'4" x 6'8" or 7'0"	Each	59	196
	3'6" x 6'8" or 7'0"	Each	66	220
	4'0" x 6'8" or 7'0"	Each	73	230
	Add for A Label	Each	8	15
	Add for B or C Label	Each	7	10
	Add for Galvanizing	Each	-	41

Labor Above Includes Installation of Butts & Locksets

Add to All Above for:
	Door Closures - Surface Mounted	Each	47	93
	Panic Bars (Exit Device)	Each	94	440
	Kick, Push and Pull Plates	Each	19	31

See 8-13 for Other Hardware

DIVISION #8 - DOORS, WINDOWS & GLASS

0802.0 WOOD DOORS (M) (CARPENTERS)
See 0603.3 and 0802.5 for Custom Frames
Labor includes Installation of Butts and Locksets

.1 **FLUSH DOORS - Pre-Machined**
No Label - Standard - Birch - 7 Ply - Paint Grade

	UNIT	LABOR	Hollow Core	Solid Core
2'0" x 6'8" or 1 3/8"	Each	47	93	103
2'4" x 6'8" or 1 3/8"	Each	48	98	108
2'6" x 6'8" or 1 3/8"	Each	49	103	113
2'8" x 6'8" or 1 3/8"	Each	50	108	118
3'0" x 6'8" or 1 3/8"	Each	52	113	124
2'0" x 6'8" or 1 3/4"	Each	48	98	113
2'4" x 6'8" or 1 3/4"	Each	49	103	118
2'6" x 6'8" or 1 3/4"	Each	50	108	124
2'8" x 6'8" or 1 3/4"	Each	51	108	129
3'0" x 6'8" or 1 3/4"	Each	54	113	134
3'4" x 6'8" or 1 3/4"	Each	58	129	149
3'6" x 6'8" or 1 3/4"	Each	62	191	220
3'8" x 6'8" or 1 3/4"	Each	68	210	230
Add for Stain Grade	Each			26
Add for 5 Ply	Each			67
Add for Jamb & Trim - Solid - Knock Down	Each			72
Add for Jamb & Trim - Veneer - Knock Down	Each			62
Add for Red Oak - Rotary	Each			21
Add for Red Oak - Plain Sliced	Each			36
Add for Architectural Grade - 7 Ply	Each			77
Deduct for Lauan	Each			15
Add for Doors over 6'8" per Inch	Each			12
Add for Vinyl Overlay	Each			46
Add for Cutouts and Special Cuts	Each			52
Add for Lite Cutouts w/Metal Frames	Each			77
Add for Private Door Eye	Each			21
Add for Wood Louvre	Each			98
Add for Transom Panels & Side Panels	Each	24		108
Add for Astragals	Each	25		41
Add for Trimming Door for Carpet	Each	20		-

			MATERIAL		
Label - Birch - Paint Grade	UNIT	LABOR	20-Min	45-Min	60-Min
2'0" x 6'8" x 1 3/4"	Each	47	124	155	160
2'4" x 6'8" x 1 3/4"	Each	48	134	165	170
2'6" x 6'8" x 1 3/4"	Each	49	139	175	180
2'8" x 6'8" x 1 3/4"	Each	50	144	210	220
3'0" x 6'8" x 1 3/4"	Each	51	149	250	260
3'4" x 6'8" x 1 3/4"	Each	57	165	270	270
3'6" x 6'8" x 1 3/4"	Each	62	230	280	280
3'8" x 6'8" x 1 3/4"	Each	68	250	290	300

See Above 0602.1 for Changes

.2 **PANEL DOORS**

			MATERIAL		
	UNIT	LABOR	Pine	Fir	Oak
Exterior Entrances					
2'8" x 6'8" x 1 3/4"	Each	73	310	320	490
3'0" x 6'8" x 1 3/4"	Each	78	330	340	520
Add for Sidelights	Each	35	180	200	320
Interior					
2'6" x 6'8" x 1 3/8"	Each	55	270	250	400
2'8" x 6'8" x 1 3/8"	Each	60	280	260	410
3'0" x 6'8" x 1 3/8"	Each	65	310	270	430

DIVISION #8 - DOORS, WINDOWS & GLASS

0802.0 WOOD DOORS (Cont'd...)

.3 LOUVERED DOORS

	UNIT	LABOR	MATERIAL Birch/Oak	Pine	Lauan
1'0" x 6'8" x 1 3/8" Pine	Each	31	360	129	-
1'3" x 6'8" x 1 3/8"	Each	34	380	134	-
1'6" x 6'8" x 1 3/8"	Each	38	400	139	-
2'0" x 6'8" x 1 3/8"	Each	47	470	165	-
2'4" x 6'8" x 1 3/8"	Each	52	470	180	-
2'6" x 6'8" x 1 3/8"	Each	52	480	220	-
2'8" x 6'8" x 1 3/8"	Each	54	490	230	-
3'0" x 6'8" x 1 3/8"	Each	64	500	250	-
Add for Half Louvered/Panel	Each				15%
Add for Prehung	Each	-	93	93	0

.4 BI-FOLD - FLUSH

	UNIT	LABOR	Birch/Oak	Pine	Lauan
2'0" x 6'8" x 1 3/8" Pine	Each	47	-	-	77
2'4" x 6'8" x 1 3/8"	Each	50	-	-	82
2'6" x 6'8" x 1 3/8"	Each	54	-	-	82
2'8" x 6'8" x 1 3/8"	Each	56	-	-	88
3'0" x 6'8" x 1 3/8"	Each	61	-	-	93
Add for Oak	Each	-	36	-	-
Add for Birch	Each	-	31	-	-
Add for Louvered Type	Each	10			82
Add for Half Louvered/Panel	Each	10			82
Add for Decorative Type	Each	10			170
Add for Prefinished Type	Each	-			21
Add for Heights over 6'8"	Each	10			4
Add for Widths over 3'0"	Each	10			5

See 0803.2 for Closet Bi-fold Doors and Prehung Stock

.5 CAFE DOORS

	UNIT	LABOR	Oak	Pine	Fir
2'6" x 3'8" x 1 1/8"	Pair	62	-	230	230
2'8" x 3'8" x 1 1/8"	Pair	68	-	240	240
3'0" x 3'8" x 1 1/8"	Pair	73	-	250	250

.6 FRENCH - with Grille Pattern

	UNIT	LABOR	Oak	Pine	Fir
2'6" x 6'8" x 1 3/8"	Each	73	440	510	310
2'8" x 6'8" x 1 3/8"	Each	75	450	520	340
3'0" x 6'8" x 1 3/8"	Each	78	460	530	460
Add for Prehung	Each	-	93	-	-

.7 DUTCH

	UNIT	LABOR	Oak	Pine	Fir
2'6" x 6'8" x 1 3/4"	Each	94	-	610	-
2'8" x 6'8" x 1 3/4"	Each	99	-	670	-
3'0" x 6'8" x 1 3/4"	Each	104	-	700	-

.8 PREHUNG DOOR UNITS - Incl. Frame, Trim, Hardware and Jambs

Exterior

	UNIT	LABOR	Birch	Pine	Fir
2'8" x 6'8" x 1 3/4"	Each	62	-	280	-
3'0" x 6'8" x 1 3/4"	Each	68	-	320	-
3'4" x 6'8" x 1 3/4"	Each	73	-	420	-
Add for Insulated	Each	10	-	-	134
Add for Threshold	Each	10	-	-	15
Add for Weatherstrip	Each	-	-	-	82
Add - Sidelites 1'0" HC	Each	21	-	-	144
Add - Sidelites 1'6" HC	Each	26	-	-	175

Interior

	UNIT	LABOR	Birch	Lauan	Oak
2'6" x 6'8" x 1 3/8"	Each	58	191	134	196
2'8" x 6'8" x 1 3/8"	Each	62	200	139	200
3'0" x 6'8" x 1 3/8"	Each	68	210	144	210

DIVISION #8 - DOORS, WINDOWS & GLASS

0803.0 SPECIAL DOORS

.1 BLAST OR NUCLEAR RESISTANT

.2 CLOSET - BI-FOLDING (PREHUNG STOCK)

.21 Wood - Units - 1 3/8"
Incl. Jamb, Csg. Track & Hdwe.

	UNIT	LABOR	Oak Flush	Birch Flush	Lauan Flush	Pine & Fir Panel
2 Door - 2'0" x 6'8"	Each	42	155	170	113	370
2'6" x 6'8"	Each	46	163	180	118	390
3'0" x 6'8"	Each	50	170	185	124	430
4 Door - 4'0" x 6'8"	Each	55	240	250	165	620
5'0" x 6'8"	Each	62	250	270	175	680
6'0" x 6'8"	Each	70	270	290	196	760

Incl. Track & Hdwe (Unfinished)

	UNIT	LABOR	Oak Flush	Birch Flush	Lauan Flush	Pine & Fir Panel
2 Door - 2'0" x 6'8"	Each	27	93	88	88	-
2'6" x 6'8"	Each	40	103	93	93	-
3'0" x 6'8"	Each	42	113	103	103	-
4 Door - 4'0" x 6'8"	Each	45	149	134	134	-
5'0" x 6'8"	Each	48	160	160	149	-
6'0" x 6'8"	Each	52	180	180	160	-
Add for Louvered Type	Each					100%

.22 Metal

	UNIT	LABOR	MATERIAL
Flush 2 Door - 2'0" x 6'8"	Each	34	77
2'6" x 6'8"	Each	36	82
3'0" x 6'8"	Each	38	88
Flush 4 Door - 4'0" x 6'8"	Each	42	113
5'0" x 6'8"	Each	44	118
6'0" x 6'8"	Each	50	129
Add for Plastic Overlay	Each	-	31
Add for Louvered or Decorative	Each	-	10

.23 Leaded Mirror

	UNIT	LABOR	MATERIAL
4 Panel Unit - 4'0" x 6'8"	Each	78	556
5'0" x 6'8"	Each	83	597
6'0" x 6'8"	Each	94	654

.3 FLEXIBLE DOORS (M) (CARPENTER)
(See 1020 - Retractable Partitions)

	UNIT	LABOR	MATERIAL
Wood Slat (Unfinished)	SqFt	2.00	10.00
Fabric Accordion Fold	SqFt	2.00	9.00
Vinyl Clad	SqFt	2.00	12.00

.4 GLASS-ALL (L&M) (GLAZIERS)

	UNIT	LABOR	COST
Manual (Including Hardware)	Each	-	2,000
Add for Center Locking	Each	-	170
Automatic (Including Hardware & Operator)			
Mat or Floor Unit	Each	-	5,800
Handle Unit	Each	-	6,300

.5 HANGAR (L&M) (IRONWORKERS)
Biparting or Sliding - Steel

	UNIT	LABOR	COST
Example: 100' x 20' ($22 SqFt)	Each	-	51,500
Add for Pass Doors	Each	-	1,200
Add for Electric Operator	Each	-	1,600
Add for Insulating	Each	-	1,800
Tilt Up or Canopy	Each	Approx. Same	

.6 METAL COVERED FIRE & INDUSTRIAL SLIDING DOORS

	UNIT	LABOR	COST
3'0" x 7'0" - Flush	Each	-	600
4'0" x 7'0"	Each	-	1,700
8'0" x 8'0" - U.L. Label	Each	-	2,300
10'0" x 10'0"	Each	-	2,900
10'0" x 12'0"	Each	-	3,200
12'0" x 12'0"	Each	-	3,600

DIVISION #8 - DOORS, WINDOWS & GLASS

0803.0 SPECIAL DOORS, Cont'd...

.7 OVERHEAD DOORS (L&M) (CARPENTERS)

	UNIT	COST 1 3/4" Wood	COST 2" 24-Ga Steel
Commercial			
8' x 8'	Each	760.00	770.00
8' x 10'	Each	890.00	910.00
8' x 12'	Each	1,000.00	1020.00
10' x 10'	Each	1,130.00	1,190.00
10' x 12'	Each	1,170.00	1,220.00
12' x 12'	Each	1,320.00	1,340.00
14' x 10'	Each	1,380.00	1,390.00
14' x 12'	Each	1,680.00	1,690.00
Add for Panel Door	Each		5%
Add for Low Lift	Each		77.00
Add for High Lift	Each		129.00
Add for Electric Operator	Each		520.00
Add for Insulated Type	Each		10%
Residential			
8' x 7' x 1 3/4"	Each	420.00	430.00
9' x 7' x 1 3/4"	Each	450.00	460.00
16' x 7' x 1 3/4"	Each	890.00	1,000.00
Add for Wood Paneled Door	Each		10%
Add for Redwood	Each		196.00
Add for Electric Operator	Each		390.00
Add for Vision Lights	Each		21.00

.8 PLASTIC LAMINATE FACED (M) (CARPENTERS)

	UNIT	LABOR	MATERIAL
1 3/8"	SqFt	4.00	11.00
1 3/4"	SqFt	4.00	11.00
Add for Edge Strip - Vertical	Each		7.00
Add for Edge Strip - Horizontal	Each		6.00
Add for Cutouts	Each		58.00
Add for Metal Frame Cutout	Each		85.00
Add for Decorative Type	SqFt		18.00
Add for Solid Colors	SqFt		10%
Add for 1-hour B Label	SqFt		5.00
Add for 1 1/2-hour B Label	SqFt		5.00
Add for Color Core	-		50%
Add for Machining for Hardware	Each		48.00

.9 REVOLVING DOOR (L&M) (GLAZIERS)

	UNIT	COST
Aluminum		
3 Leaf - 6'-6" x 6'-6"	Each	21,100.00
4 Leaf - 6'-6" x 6'-6"	Each	22,700.00
6'-0" x 7'-0"	Each	16,300.00
Automatic - 6'-6" x 10'-0"	Each	31,900.00
Stainless	Each	27,800.00
Bronze	Each	30,900.00
Painted Steel - Dark Room 32"	Each	1,700.00
36"	Each	1,800.00

DIVISION #8 - DOORS, WINDOWS & GLASS

0803.0 SPECIAL DOORS, Cont'd...   UNIT   LABOR   MATERIAL

- **.10 ROLLING DOORS & GRILLES (M) (IRONWORKERS)**
    - Doors - Manual - 8' x 8'   Each   380.00   1,750.00
        - 10' x 10'   Each   450.00   1,800.00
        - 12' x 12'   Each   440.00   2,400.00
        - 14' x 14'   Each   690.00   3,400.00
        - Add for Electric Controlled   Each   -   250.00
        - Add for Fire Doors   Each   -   590.00
    - Grilles - Manual -6'8" x 3'2"   Each   163.00   710.00
        - 6'8" x 4'2"   Each   148.00   780.00
        - 8'0" x 3'2"   Each   184.00   820.00
        - 8'0" x 4'2"   Each   194.00   900.00
        - Add for Electric Controlled   Each   -   260.00
- **.11 SHOWER DOORS (M) (CARPENTERS)**
    - 28" x 66" Obscure Pattern Glass Wire   Each   51.00   139.00
    - Tempered Glass   Each   51.00   149.00
- **.12 SLIDING OR PATIO DOORS (M) (CARPENTERS)**
    - ONE SLIDING, ONE FIXED
    - Metal (Aluminum) (Including 5/8" Glass, Threshold,
      Hardware and Screen)
        - 8'0" x 6'10"   Each   138.00   1,100.00
        - 8'0" x 8'0"   Each   184.00   1,200.00
    - Wood (M) (Carpenters) Vinyl Sheathed
      (Including 5/8" Insulated Dbl. Temp. Glass,
      Weatherstripped, Hardware and Casing)
        - Vinyl Clad 6' 0" x 6' 8"   Each   138.00   1,400.00
        - 8' 0" x 6' 8"   Each   168.00   1,700.00
        - Pine - Prefinished 6' 0" x 6' 8"   Each   138.00   1,400.00
        - 8' 0" x 6' 8"   Each   168.00   1,600.00
        - Deduct for 3/16" Safety Glass/Insulated   Each   -   170.00
        - Add for Tripe Glazing   Each   -   206.00
        - Add for Screen   Each   15.30   139.00
        - Add for Grilles   Each   18.40   206.00
- **.13 SOUND REDUCTION (M) (CARPENTERS)**
    - Metal   Each   153.00   1,500.00
    - Wood   Each   143.00   800.00
- **.14 TRAFFIC (M) (CARPENTERS)**
    - <u>Electric Powered</u>
        - Hollow Metal - 8' x 8'   Each   357.00   4,200.00
        - 8' x 8' High Speed   Each   960.00   9,300.00
        - Metal Clad (26 ga.) - 8' x 8'   Each   360.00   4,000.00
        - Wood   Each   310.00   3,300.00
    - <u>Truck Impact - Double Acting</u>
      (Including Bumpers and Vision Panels)
        - Metal Clad - 8' x 8' Openings   Each   270.00   2,300.00
        - Rubber Plastic - 8' x 8' Openings 1/2"   Each   210.00   1,900.00
        - Aluminum - 8' x 8' Openings 1 3/4"   Each   230.00   2,400.00
- **.15 SPECIAL MADE & ENGINEERED INDUSTRIAL DOORS
  (L&M) (CARPENTERS)**
    - Four-Fold Garage Doors - Electric Operated   SqFt   -   98.00
- **.16 VAULT DOORS (M) (STEEL ERECTORS)**
    - 1 hour with Frame 6' 6" x 2' 8"   Each   320.00   2,000.00
    - 2 hour with Frame 6' 6" x 2' 8"   Each   320.00   2,600.00
    - 6' 6" x 3' 0"   Each   370.00   2,900.00
    - 6' 6" x 3' 4"   Each   420.00   3,200.00

DIVISION #8 - DOORS, WINDOWS & GLASS

0804.0 ENTRANCE DOORS, FRAMES & STORE FRONT CONSTRUCTION
    Prices Include Hardware & Safety Plate Glazing

		UNIT	COST
.1	**ALUMINUM**		
.11	Stock		
	Door and Frame - 3' x 7'	Each	1,400.00
	6' x 7'	Each	2,500.00
	Window or Store Front Framing - 10'	SqFt	24.70
.12	Custom		
	Door and Frame - 3' x 7'	Each	1,700.00
	6' x 7'	Each	3,200.00
	Window or Store Front Framing - 10'	SqFt	31.90
	Add for Anodized Finish		15%
.2	**BRONZE**		
	Door and Frame - 3' x 7'	Each	3,500.00
	Window or Store Front Framing	SqFt	62.00
.3	**STAINLESS STEEL**		
	Door and Frame - 3' x 7'	Each	3,000.00
	Window or Store Front Framing	SqFt	60.00
	Add for Insulated Glass	SqFt	6.40
	Add for Exit Hardware	Each	320.00

See 0803.10 for Revolving Doors
See 0803.0 for Automatic Door Opening Assemblies

**0805.0 METAL WINDOWS (M) (CARPENTERS OR STEEL ERECTORS)
(SINGLE GLAZING INCLUDED)**

All Example Sizes Below: 2'4" x 4'6"

		UNIT	LABOR	MATERIAL	or	UNIT	LABOR	MATERIAL
.1	**ALUMINUM WINDOWS**							
	Casement and Awning	Each	48.90	240.00		SqFt	4.58	22.70
	Sliding or Horizontal Rolling	Each	48.90	175.00		SqFt	4.58	16.50
	Double and Single-Hung or Vertical Sliding	Each	48.90	175.00		SqFt	4.63	16.50
	Projected	Each	44.70	220.00		SqFt	4.06	20.60
	Add for Screens					SqFt	.62	3.61
	Add for Storms					SqFt	1.30	5.80
	Add for Insulated Glass					SqFt	-	6.00
	Add for Bronzed Finish					SqFt	-	10%
.2	**ALUMINUM SASH**							
	Casement	Each	36.40	220.00		SqFt	3.54	20.10
	Sliding	Each	36.40	155.00		SqFt	3.54	14.40
	Single-Hung	Each	36.40	155.00		SqFt	3.54	14.40
	Projected	Each	36.40	240.00		SqFt	3.54	22.70
	Fixed	Each	36.40	124.00		SqFt	2.81	11.90
.3	**STEEL WINDOWS**							
	Double-Hung	Each	44.70	350.00		SqFt	4.16	34.00
	Projected	Each	44.70	290.00		SqFt	4.16	28.80
	Add for Insulated Glass					SqFt	-	5.20
.4	**STEEL SASH**							
	Casement	Each	37.40	250.00		SqFt	3.38	23.70
	Double-Hung	Each	37.40	330.00		SqFt	3.38	31.90
	Projected	Each	37.40	280.00		SqFt	3.38	27.80
	Fixed	Each	37.40	165.00		SqFt	3.38	16.50

DIVISION #8 - DOORS, WINDOWS & GLASS

0806.0 WOOD WINDOWS (M) (Carpenters)

All Units Assembled, Kiln Dry Pine, Glazed, Vinyl Clad, Weather-stripped, and No Interior Trim. <u>Top Quality</u>. Other Sizes Available. Most commonly used listed below. Deduct 10% for Medium Quality and 25% for Low Quality Woods.

	UNIT	LABOR	MATERIAL
.1 BASEMENT OR UTILITY			
Prefinished 2'8" x 1'4"	Each	35.40	108.00
2'8" x 1'8"	Each	42.60	115.00
2'8" x 2'0"	Each	44.70	124.00
Add for Double Glazing	Each	-	39.10
Add for Storm	Each	12.50	25.80
Add for Screen	Each	10.40	17.50
.2 CASEMENT OR AWNING - Operable Units			
Single Clear Tempered Glass 1'8" x 3'0"	Each	46.80	260.00
1'8" x 3'6"	Each	47.80	270.00
2'4" x 3'6"	Each	55.00	290.00
2'4" x 4'0"	Each	57.00	350.00
2'4" x 5'0"	Each	62.00	390.00
Double (2 Units) 2'10" x 3'0"	Each	68.00	450.00
3'4" x 3'6"	Each	73.00	550.00
4'0" x 3'0"	Each	68.00	470.00
4'0" x 3'6"	Each	73.00	570.00
4'0" x 5'0"	Each	79.00	680.00
4'0" x 5'8"	Each	81.00	740.00
Triple (3 Units, 1 Fixed) 6'0" x 3'0"	Each	78.00	670.00
6'0" x 4'0"	Each	83.00	780.00
6'0" x 4'6"	Each	94.00	900.00
Add for Grille Patterns - Interior - Polycarbonate	Each	-	20.60
Add for Grille Patterns - Exterior - Prefinished	Each	-	57.00
Add for High Performance Glazing	Each	-	51.50
Add for Screen - per Unit	Each	-	18.50
Add for Extension Jambs	Each	-	24.70
Add for Blinds	Each	-	82.00
CASEMENT OR AWNING - Stationary & Fixed Units			
Single 1'8" x 3'0"	Each	46.80	220.00
1'8" x 3'6"	Each	47.80	250.00
2'4" x 3'6"	Each	55.00	260.00
2'4" x 4'0"	Each	57.00	320.00
2'4" x 5'0"	Each	64.00	350.00
2'4" x 6'0"	Each	68.00	400.00
Picture 4'0" x 4'0"	Each	39.50	410.00
4'0" x 4'6"	Each	43.70	440.00
4'0" x 6'0"	Each	45.80	590.00
6'0" x 4'0"	Each	46.80	580.00
6'0" x 6'0"	Each	52.00	760.00
Add for Grille Patterns - Wood - 2'4"	Each	20.80	77.00
Add for Grille Patterns - Wood - 4'6"	Each	26.00	118.00
.3 DOUBLE HUNG (Insulating Glass with Screen)			
2'6" x 3'0"	Each	37.40	278.00
2'6" x 3'6"	Each	40.60	290.00
2'6" x 4'2"	Each	43.70	310.00
3'2" x 3'0"	Each	50.00	300.00
3'2" x 3'6"	Each	52.00	330.00
3'2" x 4'2"	Each	54.00	340.00
Deduct for Primed Units	Each	4.16	25.80
Add for Grille Patterns	Each	7.30	25.80
Add for Triple Glazing	Each	-	61.80
Deduct for Screen Unit	Each	16.60	67.00

DIVISION #8 - DOORS, WINDOWS & GLASS

0806.0 WOOD WINDOWS, Cont'd...

		UNIT	LABOR	MATERIAL
.4	GLIDER w/SCREEN & INSUL GLASS 4'0" x 3'6"	Each	52.00	770.00
	5'0" x 3'6"	Each	57.00	830.00
	4'0" x 4'0"	Each	68.00	800.00
	5'0" x 4'0"	Each	73.00	870.00
.5	PICTURE WINDOW w/CASEMENTS (2) & INSULATING GLASS 9'6" x 5'0"	Each	130.00	1,600.00
	10'1" x 5'0"	Each	135.00	1,700.00
	9'6" x 5'6"	Each	130.00	1,700.00
	SINGLE GLASS AND STORM 10'3" x 4'6"	Each	114.00	1,700.00
	10'3" x 5'6"	Each	130.00	1,700.00
	10'3" x 6'6"	Each	135.00	1,800.00
	Add for High Performance Glass	Each	-	155.00
.6	CASEMENT ANGLE BAY WINDOWS, INSULATING GLASS, VINYL SHEATHED			
	30° (3 units) 5'10" x 4'2" x 2 7/8"	Each	182.00	1,300.00
	(4 units) 7'10" x 4'2" x 2 7/8"	Each	187.00	1,600.00
	(3 units) 5'10" x 5'2" x 2 7/8"	Each	187.00	1,400.00
	(4 units) 7'10" x 5'2" x 2 7/8"	Each	187.00	1,600.00
	(5 units) 9'10" x 4'2" x 2 7/8"	Each	220.00	1,850.00
	(5 units) 9'10" x 5'2" x 2 7/8"	Each	230.00	2,000.00
	45° (3 units) 5'4" x 4'2" x 2 7/8"	Each	192.00	1,300.00
	(4 units) 7'4" x 4'2" x 2 7/8"	Each	187.00	1,400.00
	(3 units) 5'4" x 5'2" x 2 7/8"	Each	192.00	1,400.00
	(4 units) 7'4" x 5'2" x 2 7/8"	Each	198.00	1,600.00
	(5 units) 9'4" x 4'2" x 2 7/8"	Each	210.00	1,900.00
	(5 units) 9'4" x 5'2" x 2 7/8"	Each	230.00	2,100.00
	Add for Screens	Each	41.60	46.40
	Add for Combination Windows	Each	-	240.00
	Deduct for Triple Glazing	Each	-	250.00
	Add for High Performance Glazing	Each	-	67.00
	Add for Bronze Glazing	Each	-	144.00
	Add for Color	Each	-	67.00
	Add for Blinds	Each	-	310.00
	Add for Bottom or Roof Skirt - Plywood	Each	88.00	360.00
	Add for Bottom or Roof Skirt - Aluminum	Each	94.00	540.00
.7	CASEMENT BOW WINDOWS, INSULATING GLASS			
	(3 units) 6'2" x 4'2"	Each	270.00	1,100.00
	(4 units) 8'2" x 4'2"	Each	330.00	1,500.00
	(3 units) 6'2" x 5'2"	Each	280.00	1,400.00
	(4 units) 8'2" x 5'2"	Each	350.00	1,700.00
.8	90° CASEMENT BOX BAY WINDOWS, INSULATING GLASS			
	4'8" x 4'2"	Each	350.00	1,600.00
	6'8" x 4'2"	Each	430.00	2,000.00
	4'8" x 5'2"	Each	360.00	1,600.00
	6'8" x 5'2"	Each	460.00	2,100.00
	Add for Screens - per Unit	Each	41.60	67.00
	Deduct for Double Glazing - per Unit	Each	-	36.10
	Add for High Performance Glazing	Each	-	72.10

		UNIT	LABOR	STATIONARY	VENT
.9	ROOF WINDOWS - INSULATING GLASS				
	1'10" x 3'10"	Each	166.00	370.00	590.00
	2'4" x 3'10"	Each	177.00	440.00	670.00
	3'8" x 3'10"	Each	192.00	540.00	790.00
	Add for Shingle Flashing	Each	36.40	77.00	72.00
	Add for High Performance Glass	Each	-	82.00	77.00

		UNIT	LABOR	MATERIAL
.10	CIRCLE TOPS - 2'4"	Each	99.00	490.00
	4'0"	Each	109.00	640.00
	4'8"	Each	114.00	750.00
	6'0"	Each	125.00	1,170.00

DIVISION #8 - DOORS, WINDOWS & GLASS

			UNIT	LABOR	MATERIAL
0807.0	**SPECIAL WINDOWS (M) (Carpenters/ Steel Workers)**				
.1	LIGHT-PROOF WINDOWS		SqFt	4.99	27.80
.2	PASS WINDOWS		SqFt	4.26	20.60
.3	DETENTION WINDOWS				
	Aluminum		SqFt	4.42	27.80
	Steel		SqFt	4.32	26.80
.4	VENETIAN BLIND WINDOWS (ALUMINUM)		SqFt	4.00	23.70
.5	SOUND-CONTROL WINDOWS		Each	3.80	17.50
0808.0	**DOOR AND WINDOW ACCESSORIES**				
.1	STORMS AND SCREENS (CARPENTERS)				
	Windows				
	Screen Only - Wood - 3' x 5'		Each	15.10	68.00
	Aluminum - 3' x 5'		Each	16.10	82.00
	Storm & Screen Combination - Aluminum		Each	18.20	98.00
	Doors				
	Screen Only - Wood - 3' x 6' - 8'		Each	27.00	170.00
	Aluminum		Each	25.00	191.00
	Storm & Screen Combination - Aluminum		Each	29.10	270.00
	Wood 1 1/8"		Each	31.20	250.00
.2	DETENTION SCREENS (M) (CARPENTERS)				
	Example: 4' 0" x 7' 0"		Each	78.00	480.00
.3	DOOR OPENING ASSEMBLIES (L&M) (GLAZIERS)				
	Floor or Overhead Electric Eye Units				
	Swing - Single 3' x 7' Door - Hydraulic		Each	-	3,700.00
	Double 6' x 7' Door - Hydraulic		Each	-	5,150.00
	Sliding - Single 3' x 7' Door - Hydraulic		Each	-	4,700.00
	Double 5' x 7' Door - Hydraulic		Each	-	6,400.00
	Industrial Door - 10' x 8		Each	-	6,200.00
.4	SHUTTERS - FOLDING (CARPENTERS)				
	16" x 1 1/8" x 48"		Pair	15.60	82.00
	x 60"		Pair	17.70	93.00
	x 72"		Pair	20.80	103.00

		UNIT	COST
0809.0	**CURTAIN WALL SYSTEMS (L&M) (IRONWORKERS)**		
.1	STRUCTURAL OR VERTICAL TYPE		
	Aluminum Tube Frame, Panel, Sills and Mullions - No glazing		
	Example: 4' Panel, 7' Fixed	SqFt	41.20 to 124.00
.2	PANEL WALL OR HORIZONTAL TYPE		
	Aluminum Tube Frame, Panel, Sills and Mullions - No glazing		
	Example: 1' Panel, 2' Oper., 4' Fixed	SqFt	36.10 to 100.00
.3	ARCHITECTURAL PANELS, INSULATED		
	Included in Above Prices		
	Metal		
	Baked Enamel on Steel - 24 ga	SqFt	12.90
	on Aluminum - 24 ga	SqFt	13.40
	Porcelain Enamel on Steel - 24 ga	SqFt	14.50
	on Aluminum - 24 ga	SqFt	14.90
	Exposed Aggregate - Resin	SqFt	11.30
	Fiberglass	SqFt	15.20
	Spandrel Glass	SqFt	13.40

See 0812 for Glazing Costs to be Added to Above

DIVISION #8 - DOORS, WINDOWS & GLASS

0810.0 FINISH HARDWARE (M) (CARPENTERS)

		EACH	LABOR	MATERIAL Painted	Bronze	Chrome
.1	BUTTS					
	3" x 3"	Pair	14.40	7.90	9.00	11.60
	3 1/2" x 3 1/2"	Pair	14.90	9.20	9.20	12.70
	4" x 4"	Pair	16.00	11.60	11.60	13.20
	4 1/2" x 4 1/2"	Pair	16.50	16.00	16.00	21.50
	4" x 4" Ball Bearing	Pair	17.50	30.40	30.70	39.10
	4 1/2" x 4 1/2" Ball Bearing	Pair	18.50	34.30	35.00	41.20

		UNIT	LABOR	MATERIAL
.2	CATCHES - ROLLER	Each	13.50	13.70
.3	CLOSURES - SURFACE MOUNTED - 3' - 0" Door	Each	43.70	90.00
	3' - 4"	Each	44.70	95.00
	3' - 8"	Each	45.80	98.00
	4' - 0"	Each	51.00	106.00
	CLOSURES - CONCEALED OVERHEAD - Interior	Each	73.00	148.00
	Exterior	Each	78.00	242.00
	Add for Fusible Link - Electric	Each	18.70	152.00
	CLOSURES - FLOOR HINGES - Interior	Each	104.00	470.00
	Exterior	Each	104.00	490.00
	Add for Hold Open Feature	Each	-	74.00
	Add for Double Acting Feature	Each	-	240.00
.4	DEAD BOLT LOCK - Cylinder - Outside Key	Each	33.30	115.00
	Cylinder - Double Key	Each	38.50	126.00
	Flush - Push/ Pull	Each	20.80	20.60
.5	EXIT DEVICES (PANIC) - Surface	Each	94.00	470.00
	Mortise Lock	Each	99.00	540.00
	Concealed	Each	135.00	1,700.00
	(Handrop) Automatic	Each	109.00	1,200.00
.6	HINGES, SPRING (PAINTED) - 6" Single Acting	Each	26.00	59.00
	6" Double Acting	Each	26.00	75.00
.7	LATCHSETS - Bronze or Chrome	Each	34.30	118.00
	Stainless Steel	Each	34.30	137.00
.8	LOCKSETS - Mortise - Bronze or Chrome - H.D.	Each	31.20	175.00
	S.D.	Each	29.10	127.00
	Stainless Steel	Each	29.10	237.00
	Cylindrical - Bronze or Chrome	Each	33.30	142.00
	Stainless Steel	Each	33.30	180.00
.9	LEVER HANDICAP - Latch Set	Each	31.20	147.00
	Lock Set	Each	31.20	185.00
.10	PLATES - Kick - 8" x 34" - Aluminum	Each	18.70	18.50
	Bronze	Each	19.80	39.10
	Push - 6" x 15"- Aluminum	Each	17.70	18.50
	Bronze	Each	18.70	31.90
	Push & Pull Combination - Aluminum	Each	20.80	44.30
	Bronze	Each	20.80	74.00
.11	STOPS AND HOLDERS			
	Holder - Magnetic (No Electric)	Each	41.60	108.00
	Bumper	Each	23.90	17.50
	Overhead - Bronze, Chrome or Aluminum	Each	23.90	60.00
	Wall Stops	Each	18.70	13.40
	Floor Stops	Each	18.70	14.40
	See 0808.3 for Automatic Openers & Operators			

0811.0 WEATHERSTRIPPING (L&M) (CARPENTERS)

		UNIT	LABOR	MATERIAL
.1	ASTRAGALS - Aluminum - 1/8" x 2"	Each	18.70	12.40
	Painted Steel	Each	18.70	12.40
.2	DOORS (WOOD) - Interlocking	Each	31.20	28.80
	Spring Bronze	Each	26.00	46.40
	Add for Metal Doors	Each	26.00	7.70
.3	SWEEPS - 36" WOOD DOORS - Aluminum	Each	13.50	12.90
	Vinyl	Each	13.50	9.00
.4	THRESHOLDS - Aluminum - 4" x 1/2"	Each	17.70	14.90
	Bronze - 4" x 1/2"	Each	18.70	28.30
	5 1/2" x 1/2"	Each	19.80	33.00
.5	WINDOWS (WOOD) - Interlocking	Each	34.30	23.70
	Spring Bronze	Each	29.10	41.20

DIVISION #8 – DOORS, WINDOWS & GLASS

			UNIT	COST
0812.0	**GLASS AND GLAZING (L&M) (Glaziers)**			
.1	BEVELED GLASS – 1/4" x 1/2" Bevel		SqFt	111.00
.2	COATED – Used with Insulated Glass and Heat Absorber and Light Reflector		SqFt	25.80
.3	HEAT ABSORBING AND SOLAR REJECTING COMBINATIONS of Tinted, Coated and Clear Glass in Insulated Glass Form with 1/4" or 1/2" Air Space. Usual thickness with 1/4" air space, 9/16", 11/16" and 1 3/16". With 1/2" air space, 13/16", 15/16" and 1 /16". Factory priced. Greatly variable.		SqFt	20.60
.4	INSULATED GLASS			
		1/2" and 5/8" Welded	SqFt	14.40
		1" Clear	SqFt	16.00
		Tempered Insulated	SqFt	25.80
		Triple Insulated	SqFt	22.20
		Add for Tinting	SqFt	10%
.5	LAMINATED (SAFETY AND SOUND)			
		7/32"	SqFt	13.40
		1/4"	SqFt	16.00
		3/8"	SqFt	17.00
		1/2"	SqFt	24.70
		1 3/16", 1 1/2", 1 3/4" and 2" Bulletproof	SqFt	51.50
.6	MIRROR GLASS			
		Copper Plated Back and Polished Edges	SqFt	12.40
		Unfinished	SqFt	7.70
.7	PATTERNED OR ROUGH (OBSCURE)			
		1/8"	SqFt	6.20
		7/32"	SqFt	6.40
.8	PLASTIC GLASS (SAFETY)			
		Acrylic – 1/8" .125"	SqFt	10.60
		3/16" .187"	SqFt	12.90
		1/4" .250"	SqFt	13.40
		5/16" .312"	SqFt	15.70
		3/8" .375"	SqFt	19.60
		Polycarbonates – Add to Above	SqFt	50%
.9	PLATE – CLEAR			
		1/4" (See .4 for Tinted)	SqFt	7.00
		3/8"	SqFt	9.50
		1/2"	SqFt	14.90
		5/8"	SqFt	17.50
		3/4"	SqFt	20.20
.10	SANDBLASTED – 3/16"		SqFt	8.80
.11	SHEET GLASS			
		Light Sheet or Window Glass – Double Strength	SqFt	4.33
		Heavy Sheet or Crystal	SqFt	5.80
.12	STAINED GLASS – Abstract		SqFt	77.30
		Symbolism	SqFt	113.00
		Figures	SqFt	260.00
.13	STRUCTURAL GLASS – 1/4"		SqFt	11.30
.14	TEMPERED (SAFETY PLATE)			
		1/4"	SqFt	8.00
		3/8"	SqFt	10.30
		1/2"	SqFt	22.20
		5/8"	SqFt	25.80
.15	TINTED PLATE – LIGHT REFLECTIVE			
		1/4"	SqFt	8.50
		3/8"	SqFt	11.60
		1/2"	SqFt	20.10
.16	WIRED (SAFETY)			
		1/4" – Clear	SqFt	12.90
		Obscure	SqFt	9.30

DIVISION #8 - DOORS, WINDOWS & GLASS - QUICK ESTIMATING

The costs below are average, priced as total contractor / subcontractor costs with an overhead and fee of 10% included. Units include fasteners, tools and equipment, fringe benefits, 30% taxes and insurance on labor, and 5% General Conditions.

Code	Item	UNIT	COST
0801.0	**HOLLOW METAL** (Installation of Butts and Locksets included but no hardware material)		
.1	CUSTOM FRAMES (16 ga)		
	2'6" x 6'8" x 4 3/4"	Each	220
	2'8" x 6'8" x 4 3/4"	Each	240
	3'0" x 6'8" x 4 3/4"	Each	250
	3'4" x 6'8" x 4 3/4"	Each	270
	Add for A, B or C Label	Each	31
	Add for Side Lights	Each	124
	Add for Frames over 7'0" - Height	Each	52
	Add for Frames over 6 3/4" - Width	Each	46
	CUSTOM DOORS (1 3/8" or 1 3/4") (18 ga)		
	2'6" x 6'8" x 1 3/4"	Each	260
	2'8" x 6'8" x 1 3/4"	Each	270
	3'0" x 6'8" x 1 3/4"	Each	290
	3'4" x 6'8" x 1 3/4"	Each	290
	Add for 16 ga	Each	57
	Add for B and C Label	Each	31
	Add for Vision Panels or Lights	Each	67
.2	STOCK FRAMES (16 ga)		
	2'6" x 6'8" or 7'0" x 4 3/4"	Each	155
	2'8" x 6'8" or 7'0" x 4 3/4"	Each	160
	3'0" x 6'8" or 7'0" x 4 3/4"	Each	170
	3'4" x 6'8" or 7'0" x 4 3/4"	Each	185
	Add for A, B or C Label	Each	31
	STOCK DOORS (1 3/8" or 1 3/4" - 18 ga)		
	2'6" x 6'8" or 7'0"	Each	250
	2'8" x 6'8" or 7'0"	Each	260
	3'0" x 6'8" or 7'0"	Each	280
	3'4" x 6'8" or 7'0"	Each	300
	Add for B or C Label	Each	26

Code	Item	UNIT	Hollow Core	Solid Core
0802.0	**WOOD DOORS** (Installation of Butts and Locksets included but no hardware material)			
.1	FLUSH DOORS - No Label - Paint Grade			
	Birch - 7 Ply 2'4" x 6'8" x 1 3/8"	Each	165	180
	2'6" x 6'8" x 1 3/8"	Each	170	185
	2'8" x 6'8" x 1 3/8"	Each	175	196
	3'0" x 6'8" x 1 3/8"	Each	180	210
	2'4" x 6'8" x 1 3/4"	Each	175	196
	2'6" x 6'8" x 1 3/4"	Each	180	200
	2'8" x 6'8" x 1 3/4"	Each	185	220
	3'0" x 6'8" x 1 3/4"	Each	196	220
	3'4" x 6'8" x 1 3/4"	Each	201	240
	3'6" x 6'8" x 1 3/4"	Each	211	330
	Add for Stain Grade	Each		31
	Add for 5 Ply	Each		67
	Add for Jamb & Trim - Solid - Knock Down	Each		72
	Add for Jamb & Trim - Veneer - Knock Down	Each		46
	Add for Red Oak - Rotary Cut	Each		26
	Add for Red Oak - Plain Sliced	Each		36
	Deduct for Lauan	Each		21
	Add for Architectural Grade - 7 Ply	Each		72
	Add for Vinyl Overlay	Each		52
	Add for 7'-0" Doors	Each		21
	Add for Lite Cutouts w/Metal Frame	Each		82
	Add for Wood Louvres	Each		113
	Add for Transom Panels & Side Panels	Each		139

DIVISION #8 - DOORS, WINDOWS & GLASS - QUICK ESTIMATING

0802.0 WOOD DOORS, Cont'd...

.1 FLUSH DOORS, Cont'd...
Label - 1 3/4" - Paint Grade

	UNIT	20-Min	45-Min	60-Min
2'6" x 6'8" - Birch	Each	220	250	260
2'8" x 6'8"	Each	230	260	270
3'0" x 6'8"	Each	240	280	280
3'4" x 6'8"	Each	260	300	300
3'6" x 6'8"	Each	310	310	320

.2 PANEL DOORS
Exterior - 1 3/4"

	UNIT	Pine	Fir	Birch/Oak	Lauan
2'8" x 6'8"	Each	450	460	650	-
3'0" x 6'8"	Each	480	490	680	-

Interior - 1 3/8"

	UNIT	Pine	Fir	Birch/Oak	Lauan
2'6" x 6'8"	Each	390	360	510	-
2'8" x 6'8"	Each	400	370	530	-
3'0" x 6'8"	Each	430	380	560	-
Add for Sidelights	Each	250	380	410	

.3 LOUVERED DOORS - 1 3/8"

	UNIT	Pine	Fir	Birch/Oak	Lauan
2'0" x 6'8"	Each	260	260	560	-
2'6" x 6'8"	Each	310	310	580	-
2'8" x 6'8"	Each	320	320	590	-
3'0" x 6'8"	Each	330	330	610	-

.4 BI-FOLD - 2 DOOR w/ HDWE - 1 3/8"

	UNIT	Pine	Fir	Birch/Oak	Lauan
2'0" x 6'8" - Flush	Each	-	-	200	170
2'4" x 6'8"	Each	-	-	210	180
2'6" x 6'8"	Each	-	-	220	191
Add for Prefinished	Each				21

See 0804.22 for 1 1/8" Closet Bi-folds

.5 CAFE DOORS - 1 1/8" - Pair

	UNIT	Pine	Fir	Birch/Oak	Lauan
2'6" x 3'8"	Each	370	-	-	-
2'8" x 3'8"	Each	380	-	-	-
3'0" x 3'8"	Each	400	-	-	-

.6 FRENCH DOORS - 1 3/8"

	UNIT	Pine	Fir	Birch/Oak	Lauan
2'6" x 3'8"	Each	610	480	660	-
2'8" x 3'8"	Each	620	490	680	-
3'0" x 3'8"	Each	640	500	690	-

.7 DUTCH DOORS

	UNIT	Pine	Fir	Birch/Oak	Lauan
2'6" x 3'8"	Each	780	-	-	-
2'8" x 3'8"	Each	840	-	-	-
3'0" x 3'8"	Each	890	-	-	-

.8 PREHUNG DOOR UNITS (INCL. FRAME, TRIM, HARDWARE, and SPLIT JAMBS)

Exterior Entrance - 1 3/4"

	UNIT	Pine	Fir	Birch/Oak	Lauan
Panel - 2'8" x 6'8"	Each	430	-	-	-
3'0" x 6'8"	Each	460	-	-	-
3'0" x 7'0"	Each	490	-	-	-
Add for Insulation	Each	180			
Add for Side Lights	Each	190			

Interior - 1 3/8"

	UNIT	Pine	Fir	Birch/Oak	Lauan
Flush, H.C. - 2'6" x 6'8"	Each	-	-	310	230
2'8" x 6'8"	Each	-	-	320	230
3'0" x 6'8"	Each	-	-	340	240
Add for Int. Panel Door	Each				210

DIVISION #8 - DOORS, WINDOWS & GLASS - QUICK ESTIMATING

0803.0 SPECIAL DOORS

.2 CLOSET BI-FOLDING - PREHUNG STOCK AND UNFINISHED

	UNIT	Oak Flush	Birch Flush	Lauan Flush	Pine Panel
Wood - 2 Door - 2'0" x 6'8" x 1 3/8"	Each	230	270	200	450
2'6" x 6'8" x 1 3/8"	Each	250	290	210	470
3'0" x 6'8" x 1 3/8"	Each	270	340	220	530
4 Door - 4'0" x 6'8" x 1 3/8"	Each	310	350	270	730
5'0" x 6'8" x 1 3/8"	Each	320	370	310	820
6'0" x 6'8" x 1 3/8"	Each	360	380	320	900

	UNIT	COST
Add for Prefinished -	Each	$30.90
Add for Jambs and Casings -	Each	$67.00
Metal - 2 Door - 2'0" x 6'8"	Each	139
2'6" x 6'8"	Each	144
3'0" x 6'8"	Each	165
4 Door - 4'0" x 6'8"	Each	191
5'0" x 6'8"	Each	200
6'0" x 6'8"	Each	220
Add for Plastic Overlay	Each	36
Add for Louvre or Decorative Type	Each	15
Leaded Mirror (Based on 2-Panel Unit)		
4'0" x 6'8"	Each	700
5'0" x 6'8"	Each	750
6'0" x 6'8"	Each	810
.3 FLEXIBLE DOORS (M) (CARPENTERS)		
Wood Slat (Unfinished)	SqFt	1,400
Fabric Accordion Fold	SqFt	1,300
Vinyl Clad	SqFt	1,750
.8 PLASTIC LAMINATE FACED (M) (CARPENTERS) - 1 3/8"	SqFt	1,800
1 3/4"	SqFt	1,800
Add for Decorative Type	SqFt	1,900
Add for Label Doors	SqFt	5.00
Add for Edge Strip - Vertical and Horizontal	Each	9.00
Add for Machining for Hardware	Each	4,600
.10 ROLLING - DOORS & GRILLES		
Doors - 8' x 8'	Each	2,400
10' x 10'	Each	2,600
Grilles - 6'-8" x 3'-2"	Each	900
6'-8" x 4'-2"	Each	1,000
.11 SHOWER DOORS - 28" x 66" (CARPENTERS)	Each	230
.12 SLIDING OR PATIO DOORS (M) (CARPENTERS)		
Metal (Aluminum) - Including Glass Thresholds & Screen		
Opening 8'0" x 6'8"	Each	1,500
8'0" x 8'0"	Each	1,600
Wood		
Vinyl Clad 6'0" x 6'10"	Each	1,800
8'0" x 6'10"	Each	2,000
Pine - Prefinished 6'0" x 6'10"	Each	1,800
8'0" x 6'10"	Each	2,000
Add for Grilles	Each	290
Add for Triple Glazing	Each	200
Add for Screen	Each	139
.13 SOUND REDUCTION - Metal	Each	1,750
Wood	Each	1,100
.14 TRAFFIC DOORS (CARPENTERS)		
Electric Powered - Hollow Metal - 8' x 8' Opening	Each	5,000
Wood - 8' x 8' Opening	Each	4,900
Truck Impact - Metal Clad - 8' x 8' Opening	Each	3,000
(Double Acting) - Rubber - 8' x 8' Opening	Each	2,400
Add for Frames	Each	760
.16 VAULT DOORS - 6'6" x 2'2" with Frame - 2 hour	Each	3,000

DIVISION #8 - DOORS, WINDOWS & GLASS - QUICK ESTIMATING

0805.0 METAL WINDOWS (M) (CARPENTERS OR STEEL ERECTORS) (SINGLE GLAZING INCLUDED)

All Example Sizes Below: 2'4" x 4'6"

		UNIT	COST	UNIT	COST
.1	ALUMINUM WINDOWS				
	Casement and Awning	Each	320.00	SqFt	30.90
	Sliding or Horizontal	Each	250.00	SqFt	23.70
	Double and Single-Hung or Vertical Sliding	Each	280.00	SqFt	26.80
	Projected	Each	320.00	SqFt	30.90
	Add for Screens	Each	48.40	SqFt	4.64
	Add for Storms	Each	86.50	SqFt	8.20
	Add for Insulated Glass	Each	70.00	SqFt	6.70
.2	ALUMINUM SASH				
	Casement	Each	300.00	SqFt	28.80
	Sliding	Each	250.00	SqFt	23.70
	Single-Hung	Each	250.00	SqFt	23.70
	Projected	Each	210.00	SqFt	19.60
	Fixed	Each	220.00	SqFt	20.60
.3	STEEL WINDOWS				
	Double-Hung	Each	450.00	SqFt	43.30
	Projected	Each	430.00	SqFt	41.20
.4	STEEL SASH				
	Casement	Each	340.00	SqFt	31.90
	Double-Hung	Each	390.00	SqFt	41.20
	Projected	Each	290.00	SqFt	37.10
	Fixed	Each	290.00	SqFt	27.00

0806.0 WOOD WINDOWS (M) (Carpenters) - All Units Assembled, Kiln Dry Pine, Glazed, Vinyl Clad, Weatherstripped, and Top Quality. Many Other Sizes Available.

		UNIT	COST
.1	BASEMENT OR UTILITY		
	Prefinished with Screen 2'8" x 1'4"	Each	180
	2'8" x 2'0"	Each	210
	Add for Double Glazing or Storm	Each	46
.2	CASEMENT OR AWNING		
	Operating Units - Insulating Glass		
	Single 2'4" x 4'0"	Each	400
	2'4" x 5'0"	Each	440
	Double w/Screen 4'0" x 4'0" (2 units)	Each	630
	4'0" x 5'0" (2 units)	Each	830
	Triple w/Screen 6'0" x 4'0" (3 units)	Each	990
	6'0" x 5'0" (3 units)	Each	1,100
	Fixed Units - Insulating Glass		
	Single 2'4" x 4'0"	Each	380
	2'4" x 5'0"	Each	450
	Picture 4'0" x 4'6"	Each	520
	4'0" x 6'0"	Each	680
.3	DOUBLE HUNG - & Insul. Glass w/Screen - 2'6" x 3'6"	Each	380
	2'6" x 4'2"	Each	390
	3'2" x 3'6"	Each	400
	3'2" x 4'2"	Each	460
	Add for Triple Glazing	Each	62
.4	GLIDER - & Insul. Glass w/Screen - 4'0" x 3'6"	Each	920
	5'0" x 4'0"	Each	1,100
.5	PICTURE WINDOWS w/CASEMENTS - & Insul. Glass w/Screen		
	9'6" x 4'10"	Each	1,900
	9'6" x 5'6"	Each	2,100

8A-18

DIVISION #8 - DOORS, WINDOWS & GLASS - QUICK ESTIMATING

0806.0 WOOD WINDOWS Cont'd...

.6 CASEMENT ANGLE BAY WINDOWS-Insulating Glass w/Screens

30° - 5'10" x 4'2" - (3 units)	Each	1,500
7'10" x 4'2" - (4 units)	Each	1,900
5'10" x 5'2" - (3 units)	Each	1,600
7'10" x 5'2" - (4 units)	Each	1,850
45° - 5'4" x 4'2" - (3 units)	Each	1,500
7'4" x 4'2" - (4 units)	Each	1,700
5'4" x 5'2" - (3 units)	Each	1,800
7'4" x 5'2" - (4 units)	Each	2,000

.7 CASEMENT BOW WINDOWS - Insulating Glass

6'2" x 4'2" - (3 units)	Each	1,500
8'2" x 4'2" - (4 units)	Each	2,100
6'2" x 5'2" - (3 units)	Each	1,700
8'2" x 5'2" - (4 units)	Each	2,200
Add for Triple Glazing - per unit	Each	62
Add for Bronze Glazing - per unit	Each	62
Deduct for Primed Only - per unit	Each	15
Add for Screens - per unit	Each	19

.8 90° CASEMENT BOX BAY WINDOWS - Insulating Glass

4'8" x 4'2"	Each	2,100
6'8" x 4'2"	Each	2,800
6'8" x 5'2"	Each	2,900

	UNIT	STATIONARY	MOVABLE
.9 SKY OR ROOF WINDOWS 1'10" x 3'10"	Each	580	830
2'4" x 3'10"	Each	680	980
3'8" x 3'10"	Each	810	1,200
.10 CIRCULAR TOPS & ROUNDS - 4'0"	Each	-	650
6'0"	Each	-	1,400

0807.0 SPECIAL WINDOWS (M) (CARPENTERS/ STEEL WORKERS)

.1 LIGHT-PROOF WINDOWS	SqFt	37
.2 PASS WINDOWS	SqFt	27
.3 DETENTION WINDOWS	SqFt	39
.4 VENETIAN BLIND WINDOWS (ALUMINUM)	SqFt	32
.5 SOUND-CONTROL WINDOWS	SqFt	31

0808.0 DOOR AND WINDOW ACCESSORIES

.1 STORMS AND SCREENS (CARPENTERS)

Windows

Screen Only - Wood - 3' x 5'	Each	98
Aluminum - 3' x 5'	Each	118
Storm & Screen Combination - Aluminum	Each	134

Doors

Screen Only - Wood	Each	230
Aluminum - 3' x 6' - 8'	Each	240
Storm & Screen Combination - Aluminum	Each	290
Wood 1 1/8"	Each	320

.2 DETENTION SCREENS (M) (CARPENTERS)

Example: 4' 0" x 7' 0"	Each	680

.3 DOOR OPENING ASSEMBLIES (L&M) (GLAZIERS)

Floor or Overhead Electric Eye Units

Swing - Single 3' x 7' Door - Hydraulic	Each	3,900
Double 6' x 7' Door - Hydraulic	Each	5,900
Sliding - Single 3' x 7' Door - Hydraulic	Each	5,150
Double 5' x 7' Doors- Hydraulic	Each	6,900
Industrial Doors - 10' x 8'	Each	7,000

.4 SHUTTERS (CARPENTERS)

16" x 1 1/8" x 48"	Pair	113
16" x 1 1/8" x 60"	Pair	124
16" x 1 1/8" x 72"	Pair	129

DIVISION #8 - DOORS, WINDOWS & GLASS - QUICK ESTIMATING

0810.0 FINISH HARDWARE (M) (CARPENTERS)

			UNIT	COST Painted	Bronze	Chrome
.1	BUTTS					
		3" x 3"	Pair	31	33	36
		3 1/2" x 3 1/2"	Pair	33	34	38
		4" x 4"	Pair	35	36	40
		4 1/2" x 4 1/2"	Pair	38	41	48
		4" x 4" Ball Bearing	Pair	59	61	68
		4 1/2" x 4 1/2" Ball Bearing	Pair	64	69	72

			UNIT	COST
.2	CATCHES - ROLLER		Each	36
.3	CLOSURES - SURFACE MOUNTED - 3' - 0" Door		Each	175
	3' - 4"		Each	182
	3' - 8"		Each	185
	4' - 0"		Each	198
	CLOSURES - CONCEALED OVERHEAD - Interior		Each	280
	Exterior		Each	400
	Add for Fusible Link - Electric		Each	200
	CLOSURES - FLOOR HINGES - Interior		Each	680
	Exterior		Each	710
	Add for Hold Open Feature		Each	82
	Add for Double Acting Feature		Each	270
.4	DEAD BOLT LOCK - Cylinder - Outside Key		Each	180
	Cylinder - Double Key		Each	210
	Flush - Push/ Pull		Each	57
.5	EXIT DEVICES (PANIC) - Surface		Each	680
	Mortise Lock		Each	800
	Concealed		Each	1,900
	Handicap (ADA) Automatic		Each	1,400
.6	HINGES, SPRING (PAINTED) - 6" Single Acting		Each	103
	6" Double Acting		Each	129
.7	LATCHSETS - Bronze or Chrome		Each	180
	Stainless Steel		Each	220
.8	LOCKSETS - Mortise - Bronze or Chrome - H.D.		Each	230
	Mortise - Bronze or Chrome - S.D.		Each	196
	Stainless Steel		Each	320
	Cylindrical - Bronze or Chrome		Each	210
	Stainless Steel		Each	260
.9	LEVER HANDICAP - Latch Set		Each	210
	Lock Set		Each	270
.10	PLATES - Kick - 8" x 34" - Aluminum		Each	82
	Bronze		Each	72
	Push - 6" x 15" - Aluminum		Each	46
	Bronze		Each	67
	Push & Pull Combination - Aluminum		Each	82
	Bronze		Each	118
.11	STOPS AND HOLDERS			
	Holder - Magnetic (No Electric)		Each	191
	Bumper		Each	46
	Overhead - Bronze, Chrome or Aluminum		Each	98
	Wall Stops		Each	31
	Floor Stops		Each	41
	See 0808.3 for Automatic Openers & Operators			

0811.0 WEATHERSTRIPPING (L&M) (CARPENTERS)

			UNIT	COST
.1	ASTRAGALS - Aluminum - 1/8" x 2"		Each	41
	Painted Steel		Each	41
.2	DOORS (WOOD) - Interlocking		Each	77
	Spring Bronze		Each	88
	Add for Metal Doors		Each	41
.3	SWEEPS - 36" WOOD DOORS - Aluminum		Each	36
	Vinyl		Each	26
.4	THRESHOLDS - Aluminum - 4" x 1/2"		Each	41
	Bronze 4" x 1/2"		Each	62
	5 1/2" x 1/2"		Each	67
.5	WINDOWS (WOOD) - Interlocking		Each	69
	Spring Bronze		Each	88

DIVISION #9 - FINISHES

			PAGE
0901.0	**LATH AND PLASTER (CSI 09100)**		**9-3**
.1	LATHING and FURRING		9-3
.11		Metal Lath	9-3
.12		Gypsum Lath	9-3
.13		Metal Studs	9-3
.14		Metal Trim	9-3
.2	PLASTERING		9-3
.21		Gypsum	9-3
.22		Acoustical	9-3
.23		Vermiculite	9-3
.24		Stucco or Cement Plaster	9-3
.25		Fireproofing (Sprayed On)	9-3
.26		Thin Set or Veneer	9-3
.27		Exterior Wall System	9-3
0902.0	**GYPSUM DRYWALL (CSI 09250)**		**9-4**
.1	BACKER BOARD		9-4
.2	FINISH BOARD		9-4
.3	VINYL COATED BOARD		9-4
.4	LINER PANELS		9-4
.5	METAL STUD PARTITIONS		9-4
.6	FURRING CHANNELS		9-4
.7	METAL TRIM		9-4
.8	DRYWALL TAPING and SANDING		9-5
.9	TEXTURING		9-5
0903.0	**TILEWORK (CSI 09300)**		**9-6**
.1	CERAMIC TILE		9-6
.2	QUARRY TILE		9-6
0904.0	**TERRAZZO (CSI 09400)**		**9-6**
.1	CEMENT TERRAZZO		9-6
.2	CONDUCTIVE TERRAZZO		9-6
.3	EPOXY TERRAZZO		9-6
.4	RUSTIC TERRAZZO		9-6
0905.0	**ACOUSTICAL TREATMENT (CSI 09500)**		**9-7**
.1	MINERAL TILE		9-7
.2	PLASTIC COVERED TILE		9-7
.3	METAL PAN TILE		9-7
.4	CLASS A TILE		9-7
.5	FIRE RATED DESIGN MINERAL TILE		9-7
.6	GLASS FIBRE BOARD TILE		9-7
.7	SOUND CONTROL PANELS and BAFFLES		9-7
.8	ACOUSTICAL WALL PANELS		9-7
.9	LINEAR METAL CEILINGS		9-7
0906.0	**VENEER STONE AND BRICK (Floors, Walls, Stools, Stairs)**		**9-8**
.1	LIMESTONE		9-8
.11		CUT STONE	9-8
.12		FLAGSTONE	9-8
.2	MARBLE		9-8
.3	SLATE		9-8
.4	SANDSTONE		9-8
.5	GRANITE		9-8
.6	SYNTHETIC MARBLE		9-8

DIVISION #9 - FINISHES

		PAGE
0907.0	**SPECIAL FLOORING (CSI 09700)**	**9-8**
.1	MAGNESIUM OXYCHLORIDE (Seamless)	9-8
.2	ELASTOMERIC	9-8
.3	SYNTHETIC	9-8
.4	ASPHALTIC MASTIC	9-8
.5	BRICK FLOORING	9-8
.6	PRECAST PANELS	9-8
.7	NEOPRENE	9-8
0908.0	**RESILIENT FLOORING (CSI 09650)**	**9-9**
.1	ASPHALT	9-9
.2	CORK	9-9
.3	RUBBER	9-9
.4	VINYL - COMPOSITION TILE	9-9
.5	VINYL - TILE	9-9
.6	VINYL - SHEET	9-9
0909.0	**CARPETING (CSI 09680)**	**9-9**
.1	NYLON	9-9
.2	WOOL	9-9
.3	OLEFIN	9-9
.4	CARPET TILE	9-9
.5	INDOOR-OUTDOOR	9-9
0910.0	**WOOD FLOORING (CSI 09550)**	**9-10**
.1	STRIP and PLANK FLOORING	9-10
.2	WOOD PARQUET FLOORING	9-10
.3	RADIANT TREATED FLOORING	9-10
.4	RESILIENT WOOD FLOOR SYSTEMS	9-10
.5	WOOD BLOCK INDUSTRIAL FLOORS	9-10
.6	FINISHING (SANDING, STAINING, SEALING & WAXING)	9-10
0911.0	**SPECIAL COATINGS (CSI 09800)**	**9-11**
.1	CEMENTITOUS (GLAZED CEMENT)	9-11
.2	EPOXIES	9-11
.3	POLYESTERS	9-11
.4	VINYL CHLORIDES	9-11
.5	AGGREGATE to EPOXY COATING	9-11
.6	SPRAYED FIREPROOFING	9-11
0912.0	**PAINTING (CSI 09900)**	**9-11**
0913.0	**WALL COVERING (CSI 09950)**	**9-12**
.1	VINYL PLASTICS	9-12
.2	PLASTIC WALL TILE	9-12
.3	WALLPAPER	9-12
.4	FABRICS	9-12
.5	LEATHER	9-12
.6	WOOD FLEXIBLE VENEERS	9-12
.7	FIBRE GLASS	9-12
.8	CORK	9-12
.9	GLASS	9-12
.10	GRASS CLOTH	9-12

All items in this division are subcontractors' costs and include all labor, materials, equipment and fees.

DIVISION #9 - FINISHES

0901.0 LATH & PLASTER (L&M) (Latherers and Plaster)

		UNIT	COST
.1	LATHING		
.11	Metal Lath	SqYd	13.50
.12	Gypsum Lath	SqYd	12.50
.13	Metal Studs	SqYd	14.90
.14	Channels - Studs - 16" O.C.	SqYd	14.40
	Furring - Beams & Columns - 16" O.C.	SqYd	18.10
	Ceilings (including hangers) - 16" O.C.	SqYd	20.30
.15	Metal Trim - Base Bead and Corner Bead	LnFt	3.40
	Picture Mould	LnFt	4.53
.2	PLASTERING		
.21	Gypsum (Putty Coat, Sand Float Textured), Walls - 1 Coat	SqYd	15.80
	3 Coat	SqYd	44.10
	Add for Ceiling, Column and Beam Applications	SqYd	4.20
.22	Acoustical	SqYd	25.20
.23	Vermiculite	SqYd	23.10
.24	Stucco or Cement Plaster		
	On Masonry - Large Areas - Float Finish - 1 Coat	SqYd	26.30
	3 Coat	SqYd	34.70
	Panels, Facias, Soffits - 3 Coat	SqYd	40.40
	On Metal Lath - Large Areas - 3 Coat	SqYd	46.20
	Panels, Facias, Soffits, Columns	SqYd	53.60
	On Metal Studs - 4" - 16 ga	SqYd	32.00
	6" - 16 ga	SqYd	44.10
	Deduct for 18 ga Studs	SqYd	10%
	Add for Trowel Finish	SqYd	4.52
.25	Fireproofing (Sprayed On) - Mineral Fiber		
	Steel and Decks - 1 Hour	SqYd	9.80
	2 Hour	SqYd	16.30
	Beams and Columns - Not Wrapped - 2 Hour	LnFt	17.30
	3 Hour	LnFt	20.20
.26	Thin Coat or Veneer Plaster - Walls - 1 Coat	SqYd	13.70
	2 Coat	SqYd	20.80
	Ceilings - 2 Coat	SqYd	26.30
	Add for Patching	SqYd	50%
	Add Color to Plaster and Stucco	SqYd	5.00
	Add for Scaffold - Average	SqYd	8.90
.27	Exterior Wall System - Stucco or Synthetic Plaster		
	Includes Structural Board and 2" Insulation	SqYd	99.80
	Add for Esthetic Grooving	LnFt	4.73

			FINISHED	
LATH AND GYPSUM PLASTER IN COMBINATION UNITS:		UNIT	1 Side	2 Side
WALLS				
1 Coat Plaster on Waterproofed Found. Walls		SqYd	15.90	-
2 Coat Plaster on Masonry and Concrete		SqYd	31.80	64.70
on Gypsum Lath		SqYd	39.22	89.00
3 Coat Plaster on Metal Lath		SqYd	48.80	99.60
2 Coat Plaster on Metal Studs & Gypsum Lath		SqYd	65.70	106.00
CEILINGS				
3 Coat Plaster on Metal Lath and Channel		SqYd	93.30	-
for Suspended Ceiling (including hangers)				
3 Coat Plaster and Metal Lath attached to		SqYd	77.40	-
joists, beams, etc.				
FURRED				
Beam - 3 Coat Plaster, Metal Lath, Channel		SqYd	93.30	-
Column & Pipe - 3 Coat Plaster - Metal Lath		SqYd	88.00	-
Cabinet - 3 Coat Plaster, Metal Lath, Chnl.		SqYd	59.40	-
2 Coat Plaster on Gypsum Lath		SqYd	55.10	-

DIVISION #9 - FINISHES

0902.0 GYPSUM DRYWALL (L&M) (Carpenters)
Based on 8' Ceiling Heights & Screwed-on Application

		UNIT	COST
.1	BACKER BOARD		
	3/8"	SqFt	.95
	1/2"	SqFt	.98
	5/8" Fire Rated	SqFt	.99
	Deduct for Nailed-on	SqFt	.08
.2	FINISH BOARD - Walls		
	1/2" - Standard	SqFt	1.59
	Fire Rated	SqFt	1.62
	Moisture Resistant	SqFt	1.70
	Insulating (Foil Backed)	SqFt	1.75
	Combined Fire Rated and Insulating	SqFt	1.80
	5/8" - Standard	SqFt	1.59
	Fire Rated	SqFt	1.61
	Moisture Resistant	SqFt	1.61
	Insulating (Foil Backed)	SqFt	1.64
	Combined Fire Rated and Insulating	SqFt	1.72
	Lead Lined with #4 Lead - 1/16"	SqFt	7.40
	#2 Lead - 1/16"	SqFt	5.80
	Add for Ceiling Work - to Wood	SqFt	.16
	Add for Ceiling Work - to Steel	SqFt	.40
	Add for Adhesive Method	SqFt	.17
	Deduct for Nailed On	SqFt	.08
	Add for Beam and Column Work	SqFt	.69
	Add for Resilient Clip Application	SqFt	.80
	Add for Small Cut Up Areas	SqFt	.58
	Add for Each Floor Added	SqFt	.07
	Add for Work Above 10'	SqFt	.24
	Add for Filling Hollow Metal Frames	Each	70.00
.3	VINYL COATED BOARD - 1/2"	SqFt	2.01
	5/8"	SqFt	2.12
.4	LINER PANELS - 1"	SqFt	2.65

				COST	
			UNIT	25 ga	20 ga
.5	STEEL STUDS (Incl. Top & Bottom Plates)				
	16" O.C. - 1 5/8"		SqFt	1.34	1.55
	2 1/2"		SqFt	1.36	1.57
	3 5/8"		SqFt	1.32	1.52
	6"		SqFt	1.38	1.58
	24" O.C. - 1 5/8"		SqFt	1.03	1.26
	2 1/2"		SqFt	1.07	1.29
	3 5/8"		SqFt	1.08	1.34
	6"		SqFt	1.16	1.44
	Add for Work Over 10' Heights		SqFt	.16	.15
	Add for 16 Ga 6" Galvanized		SqFt	1.60	1.18
.6	FURRING CHANNELS				
	16" O.C. 3/4" x 1 3/8" walls		SqFt	-	1.13
	24" O.C. 3/4" x 1 3/8" walls		SqFt	-	1.03
	Resilient Channels - 16" O.C.		SqFt	-	1.03
	Add for Ceiling Work		SqFt	-	.15
	Add for Beam and Column Work		SqFt	-	1.08
.7	METAL TRIM - Casing Bead		LnFt	-	1.08
	Corner Bead		LnFt	-	.84
.8	TAPING AND SANDING		SqFt	-	.49
.9	TEXTURING		SqFt	-	.57

See 0602.34 for Labor & Material Priced Separately

DIVISION #9 - FINISHES

0902.0 GYPSUM DRYWALL, Cont'd...

COMBINATION UNITS - METAL FRAMING & BOARD

Description	UNIT	COST
PARTITIONS 3 5/8" - 25 ga Studs - 24" O.C.		
1/2" Board One Side	SqFt	2.76
Two Sides	SqFt	3.87
1/2" Fire Rated Board Two Sides	SqFt	3.92
1/2" Fire Rated Board One Side - 2 Layers 2nd Side (1 Hr.)	SqFt	4.88
2 Layers Each Side (2 Hr.)	SqFt	5.80
5/8" Board One Side	SqFt	2.60
Two Sides	SqFt	4.19
5/8" Fire Rated Board Each Side (1 Hr.)	SqFt	4.24
5/8" Fire Rated Board - 2 Layers Each Side (2 Hr.)	SqFt	6.00
3 Layers to Columns & Beams (2 Hr.)	SqFt	4.51
Add for Studs 16" O.C. - 25 Ga	SqFt	.26
Deduct for 1 5/8" Studs - 25 Ga	SqFt	.23
Deduct for 2 5/8" Studs - 25 Ga	SqFt	.12
Add for Studs - 20 Ga and 16" O.C.	SqFt	.28
Add for Studs - 20 Ga and 24" O.C.	SqFt	.26
Add for 6" Studs - 25 Ga and 16" O.C.	SqFt	.12
Add for 6" Studs - 20 Ga and 16" O.C.	SqFt	.20
Add for Sound Deadening	SqFt	.82
PARTITIONS - CAVITY SHAFT WALL -		
25 Ga Studs - 24" O.C. - All Board Fire Code C		
1" Liner Board One Side - 5/8" Board 2nd Side (1 Hr.)	SqFt	5.10
2 Layers - 1/2" Board 2nd Side (2 Hr.)	SqFt	5.60
3 Layers - 5/8" Board 2nd Side (3 Hr.)	SqFt	6.70
1" Liner Board and 5/8" Board One Side		
2 Layers 1" Liner Board 2nd Side	SqFt	9.00
PARTITIONS - DEMOUNTABLE		
2 1/2" Studs 24" O.C.-1/2" Vinyl Coated Bd. 2 Sides	SqFt	4.72
3 5/8" Studs 24" O.C.-1/2" Vinyl Coated Bd. 2 Sides	SqFt	5.10
Add for 16" O.C.	SqFt	.32
Add for 5/8" Vinyl Coated Board 2 Sides	SqFt	.35
Add for Special Colors Baked On	SqFt	25%
FURRED WALLS		
16" O.C. 1/2" Board	SqFt	2.54
24" O.C. 1/2" Board	SqFt	2.60
Resilient Channels and 1/2" Board	SqFt	2.54
CEILINGS		
To Wood Joists - 1/2" Fire Rated Board (1 Hr.)	SqFt	1.33
To Steel Joists with Furring Channels		
24" O.C. - 1/2" F.R. Board (2 Hr. with 2 1/2" Conc.)	SqFt	2.49
5/8" F.R. Board (1 Hr. with 2" Conc.)	SqFt	2.54
12" O.C. - 5/8" F.R. Board (2 Hr. with 2 1/2" Conc.)	SqFt	3.07
COLUMNS - 14 WF and Up		
1/2" F.R. Board w/ Screw Studs at Corners (1 Hr.)	SqFt	3.18
2 Layers - 1/2" F.R. Bd. with Screw Studs @ Corners (2 Hr.)	SqFt	3.71
Add for Corner Bead	LnFt	1.17

See 0602.34 for Labor and Material Priced Separately
See 0703.0 for Insulation Work

DIVISION #9 - FINISHES

		UNIT	COST
0903.0	**TILEWORK (L&M) (Tile Setters)**		
.1	CERAMIC TILE		
	Floors - 1" Square - Thin Set (Unglazed)	SqFt	8.10
	— 1" Hex - Thin Set	SqFt	8.80
	1" x 2"	SqFt	8.90
	2" Square	SqFt	8.30
	16" Square	SqFt	8.80
	Add for Mortar Setting Bed	SqFt	4.68
	Add for Random Colors	SqFt	1.04
	Add for Epoxy Joints	SqFt	2.39
	Conductive - In Mortar Setting Bed	SqFt	12.50
	Walls - 2" x 2" - Thin Set	SqFt	10.00
	4 1/4" x 4 1/4"	SqFt	7.60
	4 1/4" x 6"	SqFt	7.70
	4 1/4" x 8"	SqFt	8.00
	6" x 6"	SqFt	8.40
	Add for Plaster Setting Bed	SqFt	4.58
	Add for Decorator Type	SqFt	6.50
	Base - 4 1/4"	LnFt	7.80
	6"	LnFt	7.80
	Add for Coved Base	LnFt	.94
	Accessories - Toilet and Bath Embedded	Each	57.20
.2	QUARRY TILE		
	Floors - 4" x 4" - Thin Set	SqFt	8.20
	6" x 6"	SqFt	8.00
	4" x 8"	SqFt	8.10
	Hexagon, Valencas, etc.	SqFt	9.20
	Walls - 4" x 4" - Thin Set	SqFt	8.90
	6" x 6"	SqFt	8.70
	Stairs	SqFt	11.50
	Base - 4" x 4" x 1/2"	LnFt	8.50
	6" x 6" x 1/2"	LnFt	9.50
	Add for Hydroment Joints	SqFt	.88
	Add for Epoxy Joints	SqFt	2.29
	Add for Non-Slip Tile	SqFt	.52
	Add for Mortar Setting Bed	SqFt	4.68
0904.0	**TERRAZZO (L&M) (Terrazzo Workers) - Based on Gray Cement, 4' Panel, and White Dividing Strips**		
.1	CEMENT TERRAZZO		
	Floors - 2" Bonded	SqFt	12.00
	2 1/2" Not Bonded	SqFt	10.60
	Base - 6" without Grounds or Base Bead	LnFt	19.20
	6" with Grounds and Base Bead	LnFt	23.40
	Stairs - Treads and Risers - Tread Length	LnFt	49.40
	Stringers without Base Bead	LnFt	44.20
	Pan Filled Treads and Landings	LnFt	20.40
	Add for White Cement	SqFt	.73
	Add for Brass Dividing Strips	SqFt	1.04
	Add for Large Chips (Venetian) - 3"	SqFt	5.80
	Add for 2-Color Panels	SqFt	3.43
	Add for 3-Color Panels	SqFt	4.06
	Add for Non-Slip Abrasives	SqFt	1.00
	Add for Small or Scattered Areas	SqFt	15%
.2	CONDUCTIVE TERRAZZO - 2" Bonded	SqFt	13.00
	1/4" Thin Set Epoxy	SqFt	11.80
.3	EPOXY TERRAZZO - 1/4" Thin Set	SqFt	9.90
.4	RUSTIC TERRAZZO (WASHED AGGREGATE) - Marble	SqFt	6.70
	Quartz	SqFt	7.30
	See 0907 for Precast Terrazzo Tiles		

DIVISION #9 - FINISHES

0905.0 ACOUSTICAL TREATMENT (L&M) (Carpenters)

		UNIT	COST
.1	MINERAL TILE (No Backing or Supports)		
	Fissured or Perforated - 1/2" x 12" x 12" or 12" x 24"	SqFt	2.08
	5/8" x 12" x 12" or 12" x 24"	SqFt	2.34
	3/4" x 12" x 12" or 12" x 24"	SqFt	2.50
	Add to all above for Backing or Supports if needed:		
	Ceramic Faces	SqFt	.51
	Concealed Z Spline Application	SqFt	.64
	Wood Furring Strips	SqFt	.71
	Rock Lath	SqFt	.94
	1 1/2" Channel Suspension	SqFt	1.30
	Resilient Furring Channels	SqFt	.94
.2	PLASTIC COVERED TILE (No Backing or Supports)		
	3/4" x 12" x 12"	SqFt	2.39
	5/8" x 12" x 12"	SqFt	1.98
.3	METAL PAN TILE (Supports Included)		
	12" x 24" or 12" x 36" or 12" x 48" - Steel (24 Ga)	SqFt	7.18
	Aluminum (.025)	SqFt	8.84
	Aluminum (.032)	SqFt	10.20
	Stainless Steel	SqFt	12.20
	Add for Color	SqFt	.52
.4	CLASS A MINERAL TILE		
	(Grid System w/ Suspension incl. & Partitions to Clg.)		
	5/8" x 24" x 24" or 24" x 48" - Perforated	SqFt	1.72
	Fissured	SqFt	1.72
	Add for Partitions thru Ceilings	SqFt	.26
	Add for Translucent Panels for Lights	SqFt	1.40
.5	FIRE RATED DESIGN MINERAL TILE		
	(Grid System w/ Suspension incl. & Partitions to Clg.)		
	5/8" x 24" x 24" or 24" x 48" - Perforated (2-hour rating)	SqFt	1.87
	Fissured (2-hour rating)	SqFt	1.87
	3/4" x 24" x 24" or 24" x 48" - Perforated (3-4 Hr rating)	SqFt	2.03
	Fissured (3-4 Hr rating)	SqFt	2.03
	Add for Partitions thru Ceilings	SqFt	.26
	Add for Reveal Edge or Shadow Line (24" x 24" only)	SqFt	.36
	Add for Nubby Tile	SqFt	1.30
	Add for Ceramic Faces (5/8")	SqFt	.78
	Add for Colored Tile	SqFt	.26
	Add for Fixture Fire Protection - Fixtures only	SqFt	1.09
.6	GLASS FIBRE BOARD (Grid System w/ Suspension included)		
	5/8" x 24" x 24"	SqFt	1.82
	3/4" x 24" x 48"	SqFt	1.87
	1" x 48" x 48"	SqFt	1.82
	1½" x 60" x 60"	SqFt	3.22
	Add for Three-Dimensional	SqFt	3.12
.7	SOUND CONTROL PANELS AND BAFFLES	SqFt	12.80
.8	ACOUSTICAL WALL PANELS - Special	UNIT	COST
	Fiberglass with Cloth Face - 3/4"	SqFt	4.16 to 10.40
	1"	SqFt	5.20 to 12.00
	1½"	SqFt	8.30 to 14.00
	3/4" Mineral	SqFt	1.35 to 3.72
	1/2" Fibre Board	SqFt	.88 to 1.30
.9	LINEAR METAL CEILINGS		
	4" Wide - Painted	SqFt	6.00 to 7.90
	Chrome or Brass	SqFt	8.60 to 10.90

See 0307 for Fibre Cementitious Acoustical Plank

DIVISION #9 - FINISHES

0906.0 VENEER STONE (Floors, Walls, Stools, Stairs, Partitions)
(2" and Under) (L&M) (Marble Setters)

		UNIT	COST
.1	LIMESTONE		
.11	Cut Stone - Floors 7/8" Interior	SqFt	30.50
	Floors 2" Exterior	SqFt	34.70
	Walls 1"	SqFt	33.60
	Stools 1" x 8"	LnFt	41.00
.12	Flagstone Floors 1"	SqFt	36.80
.2	MARBLE Floors 5/8"	SqFt	31.50
	Floor Tiles 3/8" x 12" x 12"	SqFt	33.60
	Walls 3/8" x 12" x 12"	SqFt	36.80
	Base 1" x 6"	LnFt	27.30
	Stool 1" x 8"	LnFt	37.80
	Stairs - Treaded Risers	SqFt	73.50
.3	SLATE Floors 3/8" Package	SqFt	20.50
	Floors 3/4" Random or Flagstone	SqFt	22.10
	Walls 3/4" Random	SqFt	27.30
	Stools 1" x 8"	LnFt	38.90
.4	SANDSTONE Floors 2"	SqFt	34.70
.5	GRANITE Floors 5/6" Interior	SqFt	41.00
	Walls - 2" Exterior	SqFt	49.40
	1 1/2" Interior	SqFt	45.20
	Counter Tops 1 1/4"	SqFt	67.20
.6	SYNTHETIC MARBLE Floors 5/8"	SqFt	18.40
	Walls 5/8"	SqFt	21.00
	Add for Leveling Old Floors, per Square Foot	1.84 to	4.00

0907.0 SPECIAL FLOORING (L&M)

		UNIT	COST
.1	MAGNESIUM OXYCHLORIDE (Seamless)	SqFt	8.80
.2	ELASTOMERIC	SqFt	10.00
.3	SYNTHETIC		
	Epoxy with Granules 1/16" Single Speed	SqFt	5.00
	3/32" Double Speed	SqFt	6.10
	Epoxy with Aggregates 1/4" Industrial	SqFt	7.90
	Polyester with Granules 3/32"	SqFt	6.60
	1/4"	SqFt	8.20
	See 0904.3 for Epoxy Terrazzo and 0603.2 for Counter Tops		
.4	ASPHALTIC MASTIC	SqFt	8.00
	Add for Acid Proof	SqFt	1.42
.5	BRICK PAVERS - Mortar Setting Bed Application		
	3 7/8" x 8" x 1 1/4" - Red	SqFt	9.70
	Iron Spot	SqFt	11.00
	Acid Proof	SqFt	21.00
	See 0206.3 for Patios and Walks on Sand Cushion		
.6	PRECAST PANELS		
	12" x 12" x 1" Precast Terrazzo	SqFt	26.30
	12" x 12" x 1" Precast Exposed Aggregate	SqFt	26.30
	12" x 12" x 1" Ceramic (Terra Cotta)	SqFt	31.50
	6" Precast Terrazzo Base	LnFt	21.00
.7	NEOPRENE 1/4"	SqFt	11.60

DIVISION #9 - FINISHES

0908.0 RESILIENT FLOORING (L&M)
(Carpet and Resilient Floor Layers)

		UNIT	COST
.1	ASPHALT - Tile 9" x 9" x 1/8"	SqFt	2.60
.2	CORK - Tile 1/8" - Many Sizes	SqFt	3.31
	3/16" - Many Sizes	SqFt	3.89
3	RUBBER		
	Tile 12" x 12" x 3/32"	SqFt	4.62
	12" x 12" x 3/32" Conductive	SqFt	5.78
	Base 2 1/2"	LnFt	1.42
	4"	LnFt	1.58
	6"	LnFt	1.79
	Treads, Stringers and Risers		
	Treads 3/16" Gauge Pre-formed	LnFt	11.00
	1/4" Gauge Pre-formed	LnFt	12.60
	5/16" Gauge Pre-formed	LnFt	16.30
	Stringers 10" x 1/8"	LnFt	8.40
	Risers	LnFt	4.41
.4	VINYL COMPOSITION TILE 12" x 12" x 3/32"	SqFt	1.52
	12" x 12" x 1/8"	SqFt	1.63
	Conductive	SqFt	5.50
	Treads 1/8"	LnFt	5.90
	3/16"	LnFt	7.40
	1/4"	LnFt	9.20
	Stringers 10" x 1/8"	LnFt	9.10
	Risers 7" x 1/8"	LnFt	4.10
	Add for Marbleized or Patterns	SqFt	1.26
.5	VINYL TILE 9" x 9" x 1/8" Solid	SqFt	6.50
	12" x 12" x 1/8"	SqFt	6.70
	Base 2 1/2"	LnFt	1.58
	4"	LnFt	1.63
	6"	LnFt	1.84
.6	VINYL SHEET 1/8" Inlaid	SqFt	2.60 – 7.00
	1/8" Homogenous	SqFt	4.31 - 7.60
	Acoustic Cushioned	SqFt	4.73 - 10.90
	Coving	LnFt	6.30
.7	VINYL CORNER GUARDS	LnFt	5.60
	Add Plywood or Masonite Underlayment	SqFt	1.68
	Add Diagonal Cuttings	SqFt	.58
	Add Clean and Wax	SqFt	.44
	Add Skim Coat Leveling	SqFt	1.94 - 3.57

0909.0 CARPETING (L&M) (Carpet Layers)

		UNIT	CUT PILE	LOOP PILE
.1	NYLON - 20 oz	SqYd		14.20 - 22.60
	22 oz	SqYd		15.20 - 23.60
	24 oz	SqYd	12.60 - 16.80	16.30 - 27.30
	26 oz	SqYd		17.30 - 27.30
	28 oz	SqYd		17.30 - 28.40
	30 oz	SqYd	15.20 - 21.00	18.90 - 31.50
	32 oz	SqYd	16.30 – 22.00	20.00 - 33.60
.2	WOOL - 36 oz	SqYd		26.30 - 63.00
	42 oz	SqYd		33.60 - 75.60
.3	OLEFIN - 20 oz	SqYd		12.60 - 20.00
	22 oz	SqYd		13.70 - 21.00
	24 oz	SqYd		15.20 - 22.00
	26 oz	SqYd		16.30 - 25.20
	28 oz	SqYd		17.90 - 26.30
.4	CARPET TILE, MODULAR-Tension Bonded	SqYd		35.70 - 63.00
	Tufted	SqYd	24.20 – 37.80	24.20 - 46.20
.5	INDOOR-OUTDOOR	SqYd		10.00 - 20.00
	Installation included in above prices			3.68 - 6.30

DIVISION #9 - FINISHES

		UNIT	COST
0909.0	**CARPETING, Cont'd...**		
	Add to Section 0909.0 above:		
	Add for Custom Carpet	SqYd	9.50 - 20.00
	Add for Cushion (Urethane, Rubber, Synthetic)	SqYd	3.89 - 6.30
	Add Bonded Cushion (Urethane Backed)	SqYd	1.37 - 1.58
	Add for Carpet Base, Top Edge Bound	LnFt	2.00 - 3.15
	Add for Skim Coat Leveling, Latex	SqFt	1.16 - 1.58
	Add for Border and Inset Work	LnFt	2.42 - 3.89
	Add for Reduction Strips	LnFt	1.84 - 2.73
0910.0	**WOOD FLOORING (L&M) (Carpenters)**		
.1	STRIP AND PLANK FLOORING		
	Oak Floors (based on Standard Grade)	UNIT	COST
	White Oak 25/32 1 1/2" Select Job Finished	SqFt	10.00
	2 1/4" Select Job Finished	SqFt	10.20
	3 1/4" Select Job Finished	SqFt	10.50
	Add for Clear Grade	SqFt	.37
	Deduct for Antique or Rustic Floors	SqFt	.47
	Maple Floors - Iron Bound (Edge Grain)		
	3/4" x 1 1/2" #2 & Better Gr Job Finished	SqFt	10.70
	3/4" x 2 1/4" #2 & Better Gr Job Finished	SqFt	11.30
	33/32 x 2 1/4" #2 & Better Gr Job Finished	SqFt	10.70
	33/32 x 2 1/4" #3 Grade Job Finished	SqFt	10.00
	Add for #1 Grade	SqFt	1.16
	Add to above for Cork Underlayment	SqFt	1.26
	Fir Floors		
	1" x 4" Vertical Grain B&B Job Finished	SqFt	5.30
	Flat Grain B&B Job Finished	SqFt	5.20
.2	WOOD PARQUET Pre Finished 5/16" Oak or Maple	SqFt	8.10
	FLOORING Pre Finished 3/4" Oak or Maple	SqFt	10.20
	Pre Finished 5/16" Teak	SqFt	11.00
	5/16" Walnut	SqFt	11.30
.3	RADIANT TREATED (Plastic and Wood) 5/16" x 12" x 12"	SqFt	12.30
.4	RESILIENT WOOD FLOOR SYSTEMS		
	Cushion Sleeper Floors (Maple)		
	33/32 2 1/4" #1 Grade Pre-Finished	SqFt	9.20
	#2 Grade Pre-Finished	SqFt	8.80
	#3 Grade Pre-Finished	SqFt	8.20
	Cork Underlayment included	SqFt	1.31
	Add for Plywood Underlayment (2 layers)	SqFt	2.00
	Fixed Sleeper Floors (Oak)		
	33/32 2 1/4" #1 Grade Pre-Finished	SqFt	9.20
	#2 Grade Pre-Finished	SqFt	9.10
	1/4" #3 Grade Pre-Finished	SqFt	9.00
.5	WOOD BLOCK INDUSTRIAL FLOORS (Creosoted) 2"	SqFt	.04
	2 1/2"	SqFt	4.57
	3"	SqFt	4.73
.6	FINISHING (Sanding, Sealing and Waxing)		
	Included in Job Finished Prices Above for New House	SqFt	1.89
	for New Gym	SqFt	1.84
	for Old Gym	SqFt	3.05
	Add to Section 0910.0 above:		
	Small Cut-up Areas	SqFt	30%
	Trim and Irregular Areas	SqFt	15%
	Polyurethane Finish with Stain	SqFt	20%
	Deduct for Residential Work	SqFt	10%

DIVISION #9 - FINISHES

0911.0 SPECIAL COATINGS

		UNIT	COST
.1	CEMENTITIOUS (Glazed Cement)	SqFt	2.47 - 4.20
.2	EPOXIES	SqFt	1.52 - 2.89
.3	POLYESTERS	SqFt	1.58 - 2.99
.4	VINYL CHLORIDE	SqFt	1.79 - 3.26
.5	AGGREGATE TO EPOXY - Blown on Quartz	SqFt	7.70
	Troweled on Quartz	SqFt	8.00

		UNIT	COST
.6	SPRAYED FIREPROOFING (with Perlite)		
	Column - 1" 2 Hour	LnFt	1.58
	1 3/8" 3 Hour	LnFt	1.89
	1 3/4" 4 Hour	LnFt	2.42
	Beams - 1 1/8" 2 Hour	LnFt	1.73
	1 1/4" 3 Hour	LnFt	1.79
	1 1/2" 4 Hour	LnFt	2.10
	Ceilings - 1/2" 1 Hour	SqFt	1.31
	5/8" 2 Hour	SqFt	1.42
	1" 3 Hour	SqFt	1.68

See 0901.25 for Other Fireproofing

0912.0 PAINTING (L&M) (Painters)

WALLS AND CEILINGS - BRUSH

	Unit	Cost
Brick 2 Coats	SqFt	.73
Concrete 2 Coats	SqFt	.66
Concrete Block - Interior 2 Coats	SqFt	.69
Exterior 2 Coats	SqFt	.71
Light Weight 2 Coats	SqFt	.74
2 Coats/Filler	SqFt	.90
Plaster 2 Coats	SqFt	.68
Sheet Rock 2 Coats	SqFt	.58
Steel Decking & Sidings 1 Coat	SqFt	.49
Stucco 2 Coats	SqFt	.81
Wood Siding & Shingles 2 Coats	SqFt	.80
Wood Paneling - Stain, Seal and Varnish	SqFt	1.51
Deduct for Roller Application	SqFt	.11
Deduct for Spray Application	SqFt	.16
Add for Scaffold Work	SqFt	.24
Add for Epoxy Paint	SqFt	.22

DOORS AND WINDOWS

	Unit	Cost
Door, Wood - 2 Coats	Each	40.60
Stain, Seal and Varnish	Each	61.00
Bi-fold Door - Stain, Seal and Varnish 48"	Each	58.00
60"	Each	69.00
72"	Each	81.00
Door Frame, Wood (incl Trim) - 2 Coats	Each	36.40
Stain, Seal & Varnish	Each	44.70
Door, Metal (Factory Primed) - 1 Coat	Each	33.30
2 Coats	Each	44.70
Door Frame, Metal (Factory Primed) - 1 Coat	Each	30.20
2 Coats	Each	33.30
Window, Wood (including Frame & Trim) - 2 Coats	Each	44.70
Add for Storm Windows or Screens	Each	22.98
Add for Shutters - Pair	Pair	65.50
Window, Steel	Each	41.60

FLOORS, EPOXY	SqFt	2.03

DIVISION #9 - FINISHES

		UNIT	COST	
0912.0	**PAINTING (L&M) (Painters)**			
	MISCELLANEOUS WOOD			
	Cabinets - Enamel 2 Coats	SqFt	.80	
	Stain and Varnish	SqFt	1.00	
	Enamel 3 Coats	SqFt	1.20	
	Stain, Seal and Varnish	SqFt	1.56	
	Wood Trim - Stain, Seal and Varnish - Large	LnFt	1.56	
	Small or Decorative	LnFt	1.82	
	MISCELLANEOUS METAL - 1 Coat Sprayed			
	Miscellaneous Iron and Small Steel	Ton	1.00	
	Structural Steel	Ton	.66	
	Structural Steel - 2 Sq.Ft. per Ln.Ft.	LnFt	.50	
	3 Sq.Ft. per Ln.Ft.	LnFt	.62	
	4 Sq.Ft. per Ln.Ft.	LnFt	.85	
	5 Sq.Ft. per Ln.Ft.	LnFt	1.06	
	Add for Brush	LnFt	40%	
	Add for 2 Coats	LnFt	75%	
	Gutters, Down Spouts and Flashings - 4"	LnFt	1.35	
	5"	LnFt	1.40	
	6"	LnFt	1.56	
	Piping - Prime and 1 Coat - 4"	LnFt	.68	
	6"	LnFt	.96	
	8"	LnFt	1.09	
	10"	LnFt	1.46	
	Radiators	SqFt	2.91	
	MISCELLANEOUS			
	Gold and Silver Leaf	SqFt	43.70	
	Parking Lines - Striping	LnFt	.31	
	Wood Preservatives	SqFt	.54	
	Waterproof Paints - 2 Coats	SqFt	.94	
	Clear - 1 Coat	SqFt	.49	
	2 Coat	SqFt	.70	
	Taping and Sanding	SqFt	.47	
	Texturing	SqFt	.52	

See 0910 for Wood Floor Finishing

			UNIT	COST	
0913.0	**WALL COVERINGS (L&M) (Painters)**				
.1	VINYL PLASTICS (54" Roll)				
	Light Weight		SqFt	1.51	- 1.89
	Medium Weight		SqFt	1.56	- 2.13
	Heavy Weight		SqFt	1.66	- 3.85
	Add for Sizing		SqFt	.19	- .31
	Add for 27" and 36" Rolls		SqFt	.16	- .34
.2	PLASTIC WALL TILE		SqFt	2.86	- 3.54
.3	WALLPAPER ($10.00-40.00 Roll, 36 S.F. to Roll)		SqFt	1.46	- 3.28
	Add or Deduct per Roll - $1.00		SqFt	.18	- .73
.4	FABRICS (Canvas, etc.)		SqFt	1.56	- 3.22
.5	LEATHER		SqFt	2.08	- 3.33
.6	WOOD VENEERS		SqFt	5.30	- 9.57
.7	FIBERGLASS		SqFt	2.18	- 3.59
.8	CORK - 3/16"		SqFt	1.82	- 3.95
.9	GLASS - 5/16"		SqFt	1.20	- 2.13
.10	GRASS CLOTH		SqFt	1.72	- 2.91

DIVISION #10 - SPECIALTIES

		PAGE
1001.0	ACCESS PANELS	10-2
1002.0	CHALKBOARDS AND TACKBOARDS (CSI 10110)	10-2
1003.0	CHUTES (Laundry and Waste) (CSI 11175)	10-2
1004.0	COAT & HAT RACKS AND CLOSET ACCESSORIES (CSI 10914 & 10916)	10-2
1005.0	CUBICLE CURTAIN TRACK & CURTAINS (CSI 10199)	10-2
1006.0	DIRECTORIES AND BULLETIN BOARDS (CSI 10410)	10-3
1007.0	DISPLAY AND TROPHY CASES (CSI 10110)	10-3
1008.0	FLOORS - ACCESS OR PEDESTAL (CSI 10270)	10-3
1009.0	FIREPLACES AND STOVES (CSI 10300)	10-3
1010.0	FIRE FIGHTING SPECIALTIES (CSI 10520)	10-3
1011.0	FLAG POLES (CSI 10350)	10-3
1012.0	FLOOR AND ROOF ACCESS DOORS (CSI 07720)	10-4
1013.0	FLOOR MATS AND FRAMES (CSI 12690)	10-4
1014.0	FOLDING GATES (CSI 10610)	10-4
1015.0	LOCKERS (CSI 10500)	10-4
1016.0	LOUVERS AND VENTS (CSI 10240 & 10235)	10-5
1017.0	PARTITIONS - WIRE MESH (CSI 10605)	10-5
1018.0	PARTITIONS AND CUBICLES - DEMOUNTABLE (CSI 10610)	10-5
1019.0	PARTITIONS - OPERABLE (CSI 10650)	10-5
1020.0	PEDESTRIAN CONTROLS - GATES & TURNSTILES (CSI 10450)	10-6
1021.0	POSTAL SPECIALTIES (CSI 10550)	10-6
1022.0	SHOWER & TUB RECEPTORS AND DOORS (CSI 1082)	10-6
1023.0	SIGNS, LETTERS AND PLAQUES (CSI 10420 & 10480)	10-6
1024.0	STORAGE SHELVING (CSI 10670)	10-6
1025.0	SUN CONTROL AND PROTECTIVE COVERS (CSI 10530)	10-6
1026.0	TELEPHONE AND SOUND BOOTHS (CSI 10570)	10-6
1027.0	TOILET AND BATH ACCESSORIES (CSI 10800)	10-6
1028.0	TOILET COMPARTMENTS (Including Shower and Dressing Rooms) (CSI 10155)	10-7
1029.0	URNS AND TRAYS (CSI 10800)	10-7
1030.0	WALL AND CORNER GUARDS (CSI 10260)	10-7

DIVISION #10 - SPECIALTIES

1001.0 ACCESS PANELS (M) (Lathers, Carpenters)

Size	UNIT	LABOR	Flush Flange	MATERIAL Drywall	UL Label
10" x 10"	Each	14.60	33.30	36.40	109.00
12" x 12"	Each	15.60	38.50	-	114.00
14" x 14"	Each	16.60	41.60	41.60	135.00
16" x 16"	Each	17.70	43.70	-	151.00
18" x 18"	Each	19.80	52.00	-	156.00
22" x 22"	Each	21.80	68.00	49.90	182.00
24" x 24"	Each	25.00	81.00	-	200.00
24" x 36"	Each	29.10	94.00	-	250.00
32" x 32"	Each	33.30	109.00	-	270.00
Add for Plaster Type					15%
Add for Locks					8.30

1002.0 CHALKBOARDS AND TACKBOARDS (Carpenters)

	UNIT	LABOR	MATERIAL
Chalkboards			
Slate - Natural 3/8" - New	SqFt	3.54	16.10
Resurfaced	SqFt	3.54	8.30
Porcelain Enameled Steel			
1/2" Foil Backed Gypsum and 24 Ga	SqFt	3.02	7.80
28 Ga	SqFt	3.07	7.30
Marker (White) Board	SqFt	2.91	10.40
Tackboard			
Cork - 1/4" on 1/4" Gypsum Backing - Washable	SqFt	1.56	6.20
Add for Colored Cork	SqFt	-	2.81
Deduct for Natural Style - Not Washable	SqFt	-	2.60
Trim (Aluminum)			
Trim-Slip & Snap (incl. Metal Grounds) 1/4#	LnFt	2.34	1.92
1/2#	LnFt	2.60	3.74
Map Rail - 3/4 lb/ft	LnFt	2.96	5.20
1/4 lb/ft	LnFt	2.60	2.91
Chalk Tray	LnFt	3.54	6.30
Average Trim Cost	SqFt	2.70	5.60
Combination Unit Costs (Average)			
Slate - 3/8" - with Tackboard and Trim	SqFt	3.74	13.30
Enameled Steel (24 Ga) w/Tackbd. & Trim, 8' x 4'	SqFt	3.54	8.00
Marker (White) Board	SqFt	3.64	8.30
Tackboard with Trim	SqFt	3.74	10.40
Add Vertical and Horizontal Sliding Units	SqFt	-	5.20
Portable Chalkboard Tackboard - 4' x 6'	Each	62.00	830.00
Add for Revolving Type		-	280.00

1003.0 CHUTES (Laundry & Rubbish)(M)(Sheet Metal Workers)

	UNIT	LABOR	MATERIAL
Laundry or Linen			
(16 Ga Aluminized Steel and 10' Floor Height)			
1 Floor Opening - 24" Dia - per Opening	Opng	210.00	890.00
2 Floors or Openings	Opng	197.00	740.00
6 Floors or Openings	Opng	187.00	690.00
Average	LnFt	20.80	78.00
Rubbish			
(16 Ga Aluminized Steel)			
1 Floor Opening - 24" Dia - per Opening	Opng	208.00	890.00
2 Floors or Openings	Opng	198.00	760.00
6 Floors or Openings	Opng	182.00	670.00
Add for Explosion Vent - per Opening	Opng	-	440.00
Add for Stainless Steel	Opng	-	370.00
Add for 30" Diameter	Opng	15.60	177.00
Deduct for Galvanized	Opng	-	78.00

1004.0 COAT/ HAT RACKS & CLOSET ACCESS. (M) (Carpenters)

	UNIT	LABOR	MATERIAL
ALUMINUM	LnFt	12.50	73.00
STAINLESS STEEL	LnFt	12.50	85.00

1005.0 CUBICLE CURTAIN TRACK & CURTAINS (M) (Carpenters)

	UNIT	LABOR	MATERIAL
	LnFt	8.30	15.50

DIVISION #10 - SPECIALTIES

1006.0 DIRECTORIES & BULLETIN BOARDS (M) (Carpenters)
(Directories For Changeable Letters)

	UNIT	LABOR	MATERIAL
Interior - 48" x 36" Aluminum	Each	135.00	920.00
Stainless Steel	Each	140.00	1,000.00
Bronze	Each	140.00	1,400.00
60" x 36" Aluminum	Each	156.00	1,100.00
Stainless Steel	Each	161.00	1,500.00
Bronze	Each	161.00	2,200.00
Exterior - 48" x 36" Aluminum	Each	182.00	1,300.00
Stainless Steel	Each	192.00	1,800.00
Bronze	Each	192.00	2,400.00
60" x 36" Aluminum	Each	192.00	1,800.00
Stainless Steel	Each	208.00	2,300.00
Bronze	Each	208.00	3,000.00
Illuminated - 72" x 48" Aluminum	Each	208.00	5,700.00
Stainless Steel	Each	310.00	8,400.00
Bronze	Each	310.00	10,100.00

1007.0 DISPLAY & TROPHY CASES (M) (Carpenters)

	UNIT	LABOR	MATERIAL
Aluminum Frame 8' x 6' x 1'6"	Each	310.00	3,300.00
4' x 6' x 1'6"	Each	187.00	1,800.00

1008.0 FLOORS - ACCESS OR PEDESTAL (L&M) (Carpenters)
(Includes Laminate Topping)

	UNIT	LABOR	MATERIAL
Aluminum	SqFt		19.20
Steel - Galvanized	SqFt		12.50
Plywood - Metal Covered	SqFt		11.40
Add for Carpeting	SqFt		3.12
Add for Stairs and Ramps	SqFt		-

1009.0 FIREPLACE AND STOVES - - -

1010.0 FIRE FIGHTING DEVICES (M) (Carpenters)

	UNIT	LABOR	MATERIAL
Extinguishers - 2½ Gal - Water Pressure	Each	22.90	77.00
Carbon Dioxide - 5 Lb	Each	23.90	177.00
10 Lb	Each	25.00	234.00
5 Lb	Each	22.90	55.00
Dry Chemical - 10 Lb	Each	23.90	68.00
20 Lb	Each	27.00	112.00
Halon - 2½ Lb	Each	22.90	120.00
5 Lb	Each	23.90	146.00
Cabinets - Enameled Steel - 18 Ga			
12" x 27" x 8" - Recessed - Portable Ext	Each	41.60	67.00
24" x 40" x 8" - Recessed - Hose Rack w/Nozzle	Each	78.00	140.00
Deduct for Surface Plastic	Each	-	20.80
Add for Aluminum	Each	-	62.00
Blanket & Cabinet - Enameled and Roll Type	Each	41.60	78.00

1011.0 FLAG POLES (M) (Carpenters)

	LABOR	MATERIAL
Aluminum Tapered - Ground 20' - 5" Butt	177.00	990.00
25' - 5½" Butt	198.00	1,400.00
30' - 6" Butt	240.00	1,500.00
35' - 7" Butt	320.00	2,000.00
40' - 8" Butt	440.00	3,300.00
50' - 10" Butt	550.00	4,400.00
60' - 12" Butt	700.00	7,700.00
Wall 15' - 4" Butt	187.00	920.00
Add for Architectural Grade		25%
Fiberglass Tapered - Deduct from Above Prices		20%
Steel Tapered - Deduct from Above Prices		20%
Steel Tube - Deduct from Above Prices		40%
Add for Concrete Bases	310.00	210.00

DIVISION #10 - SPECIALTIES

1012.0 FLOOR & ROOF ACCESS DOORS (M) (Carpenters)

	UNIT	LABOR	MATERIAL
Floor - Single Leaf 1/4" Steel			
24" x 24"	Each	88.00	550.00
30" x 30"	Each	114.00	580.00
30" x 36"	Each	140.00	670.00
36" x 36"	Each	166.00	700.00
Deduct for Recess Type	Each		31.20
Add for Aluminum	Each		83.00
Roof - Steel			
30" x 36"	Each	146.00	500.00
30" x 54"	Each	156.00	710.00
30" x 96"	Each	230.00	1,300.00
Add for Aluminum	Each		83.00
Add for Fusible Link	Each		31.20

1013.0 FLOOR MATS & FRAMES (M) (Carpenters)

	UNIT	LABOR	MATERIAL
Mats - Vinyl Link 4' x 5' x 1/2"	Each	38.50	330.00
4' x 7' x 1/2"	Each	41.60	480.00
Add for Color	SqFt		4.16
Frames - Aluminum 4' x 5'	Each	73.00	210.00
4' x 7'	Each	78.00	260.00

1014.0 FOLDING GATES (M) (Carpenters)

	UNIT	LABOR	MATERIAL
8' x 8' - with Lock Cylinder	Each	280.00	1,300.00
	or SqFt	4.42	20.80

1015.0 LOCKERS (Incl. Benches) (M) (Sheet Metal Workers)

			MATERIAL	
Single - Open Type with 6" Legs	UNIT	LABOR	60"	72"
9" x 12" x 72"	Each	19.80	109.00	114.00
9" x 15" x 72"	Each	19.80	114.00	120.00
12" x 12" x 72"	Each	20.80	127.00	130.00
12" x 15" x 72"	Each	21.80	128.00	132.00
12" x 18" x 72"	Each	22.40	130.00	133.00
15" x 15" x 72"	Each	23.40	132.00	136.00
15" x 18" x 72"	Each	26.00	135.00	140.00
15" x 21" x 72"	Each	27.00	146.00	151.00
Double Tier				
12" x 12" x 36"	Each	15.10	68.00	
12" x 15" x 36"	Each	15.60	71.00	
12" x 18" x 36"	Each	16.10	73.00	
15" x 15" x 36"	Each	16.10	76.00	
15" x 18" x 36"	Each	16.60	83.00	
15" x 21" x 36"	Each	17.20	88.00	
Box Lockers				
12" x 12" x 12"	Each	14.00	38.50	
15" x 15" x 12"	Each	14.60	39.50	
18" x 18" x 12"	Each	15.10	41.60	
12" x 12" x 24"	Each	15.10	52.00	
15" x 15" x 24"	Each	15.60	60.00	
18" x 18" x 24"	Each	16.10	62.00	
Add Flat Key Locks (Master Keyed)	Each	-	8.30	
Add for Combination Locks	Each	-	10.40	
Add for Sloped Tops	Each	7.30	10.40	
Add for Closed Base	Each	7.30	8.30	
Add for Other than Standard Colors	Each	-	15%	
Add for Benches	LnFt	8.30	26.00	
Add for Less than Banks of 4	Each	5%	10%	
Add for Recessed Latch (Quiet)	Each	-	3.12	

DIVISION #10 - SPECIALTIES

1016.0 LOUVERS AND VENTS (M) (Carpenters or Sheet Metal Workers)

LOUVRES	UNIT	LABOR	MATERIAL
Exterior			
Galvanized - 16 Ga	SqFt	13.50	37.50
Aluminum - 14 Ga	SqFt	13.50	48.90
Extruded 12 Ga	SqFt	13.50	56.00
Add for Brass, Bronze or Stainless	SqFt	-	43.70
Add for Insect Screen	SqFt	-	3.12
Interior Door or Partition - Steel	SqFt	12.50	43.70
Aluminum	SqFt	12.50	54.00
VENTS - Brick - Aluminum	Each	25.00	68.00
Block - Aluminum	Each	25.00	104.00
See Div. 15 for Mechanically Connected			

1017.0 PARTITIONS - WIRE MESH (M) (Carpenters)

PARTITIONS			
6 Ga - 2" Mesh & Channel Frame	SqFt	4.68	9.70
10 Ga - 1 1/2" Mesh - 8'	SqFt	3.64	7.80
Add for Door	Each	94.00	370.00
WINDOW GUARDS - 10 Ga with Frame	SqFt	5.20	15.60
Add for Galvanized	-	-	15%

1018.0 PARTITIONS AND CUBICLES - DEMOUNTABLE (L&M) (Carpenters)

PARTITIONS - DEMOUNTABLE - 1 9/16" x 10'		COST
Aluminum or Enameled Steel Extrusions (Acoustical)		
Gypsum Panels (Unpainted)	SqFt	12.30
Vinyl Laminated Gypsum Panels	SqFt	14.00
Steel Panels	SqFt	14.00
Wood Panels - Hard Wood	SqFt	8.80
Add for 2 3/4"	SqFt	1.56
Add per Door	Each	480.00
CUBICLES - DEMOUNTABLE (Includes Glass)		
Steel Frame and Panels (Acoustical) Burlap		
3' - 6" High	SqFt	21.80
4' - 8" High	SqFt	19.80
5' - 8" High	SqFt	17.20
7' - 0" High	SqFt	17.70
Aluminum Frame and Panels (Acoustical) Burlap		
3' - 6" High	SqFt	17.70
4' - 8" High	SqFt	16.60
5' - 8" High	SqFt	16.10
7' - 0" High	SqFt	15.60
Deduct for Non-Acoustical Type	SqFt	2.08
Add for Vinyl Laminated	SqFt	1.82
Add for Carpeted	SqFt	4.16
Add for Curved Panels	SqFt	7.30
Add per Door	Each	420.00
Add per Gate	Each	260.00

1019.0 PARTITIONS - OPERABLE (Including Moving Dividers and Operable Walls) (L&M) (Carpenters)

			MATERIAL
ACCORDION FOLD (Fabric)	SqFt	4.68	14.60
Add for Acoustical Type	SqFt	-	5.10
Add for Metal & Wood Supports	LnFt	4.42	6.00
FOLDING (Wood, Plastic Lam., Fabric Cover)	SqFt	5.70	30.20
Add for Acoustical Type	SqFt	-	5.70

DIVISION #10 - SPECIALTIES

		UNIT	LABOR	MATERIAL
1019.0	**PARTITIONS - OPERABLE, Cont'd...**			
	OPERABLE WALLS (Electrically Operated)			
	Folding - Plastic Laminated	SqFt	6.80	38.50
	Vinyl Overlaid	SqFt	6.80	36.40
	Wood (Prefinished)	SqFt	6.20	35.40
	Add for Chalk Bond	-	-	-
	Side Coiling - Wood	SqFt	-	46.80
	Metal	SqFt	-	52.00
	REMOVABLE OR PORTABLE WALLS (Acoustical)			
	4"	SqFt	-	52.00
	2 1/4"	SqFt	-	41.60
1020.0	**PEDESTRIAN CONTROLS - GATES & TURNSTILES**	-	-	-
1021.0	**POSTAL SPECIALTIES (M) (Carpenters)**			
	CHUTES - Aluminum	Per Floor	210.00	960.00
	Bronze	Per Floor	210.00	1,200.00
	RECEIVING BOX - Aluminum	Each	130.00	920.00
	Bronze	Each	130.00	1,350.00
	LETTER BOXES			
	3" x 5" Alum. w/ Cam Locks Front Load	Each	10.40	62.00
	Rear Load	Each	10.40	75.00
1022.0	**SHOWER & TUB RECEPTORS & DOORS (M) (Cement Finishers)**			
	PRECAST TERRAZZO W/ STAINLESS STEEL CAP			
	32" x 32" x 6"	Each	99.00	300.00
	36" x 36" x 6"	Each	109.00	370.00
	36" x 24" x 12"	Each	114.00	540.00
1023.0	**SIGNS, LETTERS & PLAQUES (M) (Carpenters)**			
	LETTERS			
	Cast Aluminum, Baked Enamel, Anodized Finish			
	6" x 1/2" x 2"	Each	11.40	27.00
	8" x 1/2" x 2"	Each	12.50	33.30
	12" x 1/2" x 2"	Each	13.50	62.40
	Plastic - Character	Each	2.60	3.64
	PLAQUES - 24" x 24" Cast Aluminum	Each	104.00	890.00
	24" x 24" Bronze	Each	104.00	1,000.00
	SIGNS - HANDICAP (Metal)	Each	13.50	36.40
1024.0	**STORAGE SHELVING (L&M) (Sheet Metal Workers)**			
	METAL - 18 Ga - 36" Sections 7'1" High			
	12" Deep, 6 Shelves	LnFt	14.00	37.40
	18" Deep, 6 Shelves	LnFt	15.10	44.70
	24" Deep, 6 Shelves	LnFt	17.30	52.00
	Add or Deduct per Shelf	LnFt	2.91	5.20
	Add for Closed Back and Ends	LnFt	4.68	5.20
	Add for 48' Sections	LnFt	4.78	3.64
	Deduct for 22 Ga	LnFt	.52	3.64
	PALLET RACKS - 6' x 36" - 4 Shelves	LnFt	29.10	104.00
1025.0	**SUN CONTROL & PROTECTIVE COVERS (L&M)**	SqFt	5.20	18.00
1026.0	**TELEPHONE AND SOUND BOOTHS (M) (Carpenters)**	Each	210.00	3,400.00
1027.0	**TOILET & BATH ACCESSORIES (L&M) (Plumbers)**			
	BASED ON STAINLESS STEEL	UNIT	LABOR	MATERIAL
	Curtain Rods - 60"	Each	18.70	31.20
	Medicine Cabinets - 14" x 24"	Each	31.20	83.20
	18" x 24"	Each	33.30	109.20
	Mirrors - 16" x 10"	Each	20.80	67.60
	18" x 24"	Each	22.90	83.20
	24" x 30"	Each	31.20	104.00
	Add for Shelves	-	-	27.00
	Mop Holders	Each	26.00	72.80
	Paper Dispenser - Surface Mounted	Each	20.80	31.20
	Recessed	Each	26.00	62.00
	Toilet Seat Covers - Recessed	Each	26.00	73.00

DIVISION #10 - SPECIALTIES

	UNIT	LABOR	MATERIAL
1027.0 TOILET & BATH ACCESSORIES, Cont'd...			
Purse Shelf	Each	22.90	62.00
Robe Hooks	Each	16.70	14.60
Handicap Grab Bar - 18"	Each	21.80	36.40
36"	Each	25.00	41.60
42"	Each	31.20	46.80
Set of Three	Each	68.00	94.00
Soap Dispenser - Wall Mount	Each	20.80	46.80
- with Lather Shelf	Each	22.90	94.00
Soap and Grab Bar	Each	20.80	52.00
Toilet Seat Cover Dispenser	Each	25.00	73.00
Towel Bars - 24"	Each	20.80	31.20
Towel Dispenser - Recessed	Each	46.80	420.00
- Surface Mounted	Each	41.60	135.00
and Disposal - Recessed	Each	39.50	420.00
Sanitary Napkin Disposal - Recessed	Each	23.90	62.00
Electric Hand Dryer	Each	62.00	340.00
1028.0 TOILET COMPARTMENTS (Incl. Shower & Dressing Rooms) (L&M) (Sheet Metal Workers & Plumbers)			
PARTITIONS - FLOOR BRACED			
(Door & 1 Side Panel) Baked Enameled	Each	125.00	400.00
Stainless Steel	Each	130.00	940.00
Plastic Laminate	Each	130.00	460.00
Marble	Each	210.00	750.00
Add for Extra Side Panel - Baked Enameled	Each	73.00	198.00
Plastic Laminate	Each	73.00	220.00
Marble	Each	146.00	280.00
Add for Ceiling Hung	Each	73.00	104.00
Add for Reinforcing for Handicap Bars (ADA)	Each	-	78.00
Add for Paper Holders	Each	15.60	36.40
Urinal Screens - Baked Enameled	Each	88.00	198.00
Plastic Laminate	Each	88.00	220.00
Marble	Each	130.00	350.00
Shower Partitions (No Plumbing)			
Baked Enameled (w/ Curtain & Terrazzo Base)			
32" x 32"	Each	156.00	620.00
36" x 36"	Each	156.00	560.00
Fiberglass & Fiberglass Base - 32" x 32"	Each	94.00	250.00
36" x 36"	Each	94.00	290.00
Add for Doors	Each	41.60	198.00
Tub Enclosures			
Aluminum Frame and Tempered Glass Door	Each	73.00	177.00
Chrome Frame and Tempered Glass Door	Each	73.00	240.00
1029.0 URNS AND TRAYS			
CIGARETTE URNS - SS Rectangular or Round	Each	41.60	94.00
1030.0 WALL AND CORNER GUARDS			
CORNER GUARDS			
Stainless Steel			
16 Ga 4" x 4" with Anchor	LnFt	6.20	19.80
18 Ga 4" x 4" with Anchor	LnFt	5.70	19.20
22 Ga 3½" x 3½" with Anchor	LnFt	6.20	16.10
22 Ga 3½" x 3½" with Tape	LnFt	5.70	15.10
Aluminum - Anodic with Tape	LnFt	5.70	13.50
Vinyl	LnFt	5.20	8.60
See 0502.3 for Steel			

DIVISION #10 - SPECIALTIES

DIVISION #11 - EQUIPMENT

		PAGE
1101.0	**MEDICAL CASEWORK (CSI 12342)**	**11-3**
.1	METAL	11-3
.2	PLASTIC - LAMINATE	11-3
.3	WOOD	11-3
.4	STAINLESS STEEL	11-3
1102.0	**EDUCATIONAL CASEWORK (CSI 12341)**	**11-3**
.1	METAL	11-3
.2	PLASTIC LAMINATE	11-3
.3	WOOD	11-3
1103.0	**KITCHEN AND BATH CASEWORK (CSI 11455)**	**11-3**
.1	PREFINISHED WOOD	11-3
.2	PLASTIC LAMINATE	11-3
1104.0	**MEDICAL EQUIPMENT (Hospital, Lab, Pharmacy) (CSI 11700)**	**11-3**
.1	STATIONS or SERVICE CENTERS	11-3
.2	LABORATORY TABLES and DEMONSTRATION DESKS (CSI 11600)	11-3
.3	TOPS and BOWLS (Special)	11-4
.4	FUME HOODS	11-4
.5	KEY CABINETS	11-4
.6	BLANKET and SOLUTION WARMERS	11-4
.7	ENVIRONMENTAL GROWTH CHAMBERS	11-4
.8	CHART RACKS and CHART RACK DESKS	11-4
.9	STERILIZING EQUIPMENT	11-4
.10	X-RAY and DARKROOM EQUIPMENT (CSI 11180)	11-5
.11	AUTOPSY and MORTUARY EQUIPMENT (CSI 11800)	11-5
.12	NARCOTIC SAFES	11-5
.13	REFRIGERATORS and FREEZERS (Other than Mortuary)	11-5
.14	TOTE TRAYS and UTILITY CARTS	11-5
.15	GAS TRACKS	11-5
.16	ANIMAL FENCING and CAGING	11-5
.17	DENTAL	11-5
.18	SURGICAL	11-5
.19	INCUBATORS	11-5
.20	PHYSICAL THERAPY	11-5
.21	OVERHEAD HOISTS	11-5
1105.0	**EDUCATIONAL EQUIPMENT (CSI 11300)**	**11-5**
1106.0	**RESIDENTIAL APPLIANCES (CSI 11452)**	**11-5**
.1	FREE STANDING	11-5
.2	BUILT IN	11-5
1107.0	**ATHLETIC, RECREATION & THERAPEUTIC EQUIPMENT (CSI 11480)**	**11-6**
.1	BASKETBALL BACKSTOPS	11-6
.2	GYM SEATING (Folding)	11-6
.3	SCOREBOARDS	11-6
.4	BOWLING and BILLIARDS	11-6
.5	SWIMMING POOL EQUIPMENT	11-6
.6	OUTSIDE RECREATIONAL EQUIPMENT	11-6
.7	GYMNASTIC EQUIPMENT	11-6
.8	SHOOTING RANGE EQUIPMENT (CSI 11496)	11-6
.9	HEALTH CLUB EQUIPMENT	11-6
1108.0	**FOOD SERVICE EQUIPMENT (CSI 11400)**	**11-7**
1109.0	**LAUNDRY AND DRY-CLEANING EQUIPMENT (CSI 11110)**	**11-7**
1110.0	**LIBRARY EQUIPMENT (CSI 11050)**	**11-7**

DIVISION #11 - EQUIPMENT

		PAGE

1111.0 ECCLESIASTICAL EQUIPMENT (CSI 11040) — 11-7
   .1   CROSSES, SPIRES and STEEPLES — 11-7
   .2   CONFESSIONALS — 11-7
   .3   PEWS — 11-7
   .4   BAPTISMAL FONTS — 11-7
   .5   ALTARS, LECTERNS, TABLES, etc. — 11-7

1112.0 STAGE AND THEATRICAL EQUIPMENT (CSI 11060) — 11-8
   .1   STAGE DRAPERIES, HARDWARE, CONTROLS & OTHER EQUIPMENT — 11-8
   .2   FOLDING STAGES — 11-8
   .3   REVOLVING STAGES — 11-8
   .4   FIXED AUDITORIUM SEATING — 11-8
   .5   PROJECTION SCREENS — 11-8

1113.0 BANK EQUIPMENT (CSI 11030) — 11-8
   .1   SAFES — 11-8
   .2   DRIVE-IN WINDOW UNITS — 11-8
   .3   TELLER UNITS — 11-8
   .4   NIGHT DEPOSITORIES — 11-8
   .5   SURVEILLANCE EQUIPMENT — 11-8
   .6   SAFE DEPOSIT BOX — 11-8

1114.0 DETENTION EQUIPMENT (CSI 11190) — 11-8

1115.0 PARKING EQUIPMENT (CSI 11150) — 11-8
   .1   AUTOMATIC DEVICES — 11-8
   .2   BOOTHS — 11-8
   .3   METERS and POSTS — 11-8

1116.0 LOADING DOCK EQUIPMENT (CSI 11160) — 11-8
   .1   DOCK BUMPERS — 11-8
   .2   DOCK BOARDS — 11-9
   .3   LEVELERS and ADJUSTABLE RAMPS — 11-9
   .4   SHELTERS — 11-9

1117.0 SERVICE STATION EQUIPMENT — 11-9
   .1   REELS — 11-9
   .2   PUMPS — 11-9
   .3   COMPRESSORS — 11-9
   .4   TIRE CHANGERS — 11-9

1118.0 MUSICAL EQUIPMENT (CSI 11070) — 11-9
1119.0 CHECKROOM EQUIPMENT (CSI 11090) — 11-9
1120.0 AUDIO VISUAL EQUIPMENT (CSI 11130) — 11-9
1121.0 INCINERATORS (CSI 11171) — 11-9
1122.0 PHOTO AND GRAPHIC ART EQUIPMENT (CSI 11470) — 11-9
1123.0 WASTE HANDLING EQUIPMENT (11860) — 11-9

DIVISION #11 - EQUIPMENT

1101.0 MEDICAL CASEWORK (L&M) (Carpenters)
   INCLUDING PLASTIC LAMINATE TOP, HARDWARE & FINISH

		UNIT	LABOR	MATERIAL
.1	METAL			
	Base Cabinets - 35" High x 24" Deep	LnFt	30.20	185.00
	Wall Cabinets - 24" High x 12" Deep	LnFt	28.10	144.00
	High Wall (Utility) - 84" High x 12" Deep	LnFt	31.20	149.00
.2	PLASTIC LAMINATE (Including Plastic Laminator)			
	Base Cabinets - 35" High x 24" Deep	LnFt	30.20	196.00
	Wall Cabinets - 24" High x 12" Deep	LnFt	28.10	134.00
	High Wall (Utility) - 84" High x 12" Deep	LnFt	31.20	160.00
.3	WOOD			
	Base Cabinets - 35" High x 24" Deep	LnFt	30.20	180.00
	Wall Cabinets - 24" High x 12" Deep	LnFt	28.10	134.00
	High Wall (Utility) - 84" High x 12" Deep	LnFt	32.20	155.00
.4	STAINLESS STEEL			
	Base - 35" High x 24" Deep	LnFt	31.20	216.00
	Wall - 24" High x 12" Deep	LnFt	28.10	200.00

1102.0 EDUCATIONAL CASEWORK (L&M) (Carpenters)
   Same as 1101.0 Medical Casework above.

1103.0 KITCHEN AND BATH CASEWORK (M) (Carpenters)

.1	PREFINISHED WOOD - NO TOPS			
	Base Cabinets - 35" High x 18" Deep	LnFt	30.20	155.00
	Front Only	LnFt	28.10	87.50
	with 1 Drawer	LnFt	30.20	185.00
	with 4 Drawers	LnFt	31.20	250.00
	with Lazy Susan	LnFt	30.20	196.00
	Upper Cabinets - 30" High x 12" Deep	LnFt	29.10	124.00
	18" High x 12" Deep	LnFt	28.10	98.00
	with Lazy Susan	LnFt	29.10	165.00
	High Cabinets - 84" High x 18" Deep	LnFt	31.20	250.00
	Add for Counter Tops - See 0603.2			
.2	PLASTIC LAMINATE			Add 10%

1104.0 MEDICAL EQUIPMENT (M) (Carpenters)

.1	STATION AND SERVICE CENTER EQUIPMENT			
.11	Kitchen -			
	Example: 4'6" x 4'2" Base & 30" Upper	Each	220.00	1,900.00
.12	Medical Preparation -			
	Example: 4'0" x 6'8" x 1'8"	Each	270.00	4,550.00
	(also 5' and 6' wide)			
.13	Nourishment and Ice Stations -			
	Example: 6'0" x 6'8" x 2'8"	Each	290.00	9,700.00
.14	Lavatory Unit -			
	Example: 4'7" x 1'8" x 4'0"	Each	135.00	900.00
.15	Janitor's Closet -			
	Example: 2'0" x 5'4" x 6'8"	Each	250.00	3,800.00
.16	Glove Packaging - Example	Each	135.00	990.00
.17	Linen Inspection - Example	Each	156.00	1,450.00
.2	LAB TABLES AND DEMONSTRATION DESKS	Each	145.60	2,150.00

DIVISION #11 - EQUIPMENT

1104.0 MEDICAL EQUIPMENT, Cont'd...

	UNIT	LABOR	MATERIAL
.3 TOPS AND BOWLS (M) (Carpenter or Plumber)			
Calcium Aluminum Silicate	SqFt	5.70	22.70
Corian	SqFt	5.70	43.30
Epoxy Resins (Kemstone)	SqFt	5.70	20.60
Marble - Artificial	SqFt	4.53	22.70
Plastic Laminate - 25" - 1 1/2"	SqFt	3.86	20.60
Add for Backsplash	LnFt	1.65	4.12
Soapstone	SqFt	4.43	30.90
Stainless Steel - Bowl	Each	65.00	175.00
Tops	SqFt	5.90	56.50
Add for Backsplash	LnFt	1.85	7.20
Vinyl Sheet - 25" - 1 1/2"	SqFt	2.88	12.40
Add for Backsplash	LnFt	1.55	4.10
.4 FUME HOODS (INCLUDING DUCT WORK)			
Stock - 4"	Each	390.00	4,750.00
Custom	Each	390.00	5,300.00
.5 KEY CABINETS (100 KEY)	Each	77.50	280.00
.6 BLANKET AND SOLUTION WARMERS (NO MECHANICAL OR ELECTRICAL)			
Example: 27 x 11 x 30" x 74			
Free Standing - Steam	Each	360.00	4,550.00
Electric	Each	360.00	4,750.00
Recessed - Steam	Each	410.00	4,450.00
Electric	Each	410.00	4,550.00
.7 ENVIRONMENTAL GROWTH CHAMBERS & LABS	Each	1,050.00	4,200.00
.8 CHART RACKS AND CHART RACK DESKS			
Racks	Each	67.00	280.00
Desks	Each	103.00	1,400.00

		ELECTRIC		STEAM	
		Re-cessed	Open Mtd or Mobile	Re-cessed	Open Mtd or Mobile
.9 STERILIZING EQUIPMENT (NO MECHANICAL/ELECTRICAL)	UNIT				
Dressing and Instrument -					
Example: 16" x 26"	Each	9,700	-	880	-
Dressing -					
Example: 20" x 20" x 38"	Each	19,600	-	15,500	1,600
Solution Warming/ Storage Cab.	Each	-	-	-	-
Bedpan -					
Example: 16" x 16" x 27"					
Non Pressure	Each	-	2,250	-	-
Pressure	Each	-	2,600	-	-
Laboratory (Painted)	Each	11,100	11,700	10,000	12,200
Add for Stainless Steel					
Autoclave (All Purpose)					
Portable Inst. - Non Pressure					
6" x 6" x 13"	Each	-	950	-	-
8" x 8" x 16"	Each	-	1,500	-	-
10" x 10" x 22"	Each	-	4,850	-	-
Hot Air -					
Example: 39" x 14" x 19"					
Gravity	Each	2,800	2,350	-	-
Convection	Each	4,100	4,350	-	-

DIVISION #11 - EQUIPMENT

1104.0 MEDICAL EQUIPMENT, Cont'd...
    .10   X-RAY AND DARK ROOM EQUIPMENT
         Lead Protection - See 0801 for Doors & View Window Frames
         Lead Protection - See 0902 for Lead Lined Sheetrock
         Manual Dark Room

	UNIT	LABOR	MATERIAL
Example: Room including Cassette Pass Box, Cassette Storage Cabinet, Film Identifier Cabinet, Film Illuminator and Dryer, Development Tank including Tops and Splashes	Each	1,450.00	14,400.00
Automatic			
Processor with Solution Storage Tank	Each	770.00	10,800.00
Film Loader	Each	260.00	5,400.00
.11 AUTOPSY AND MORTUARY EQUIPMENT			
Refrigerators	Each	440.00	9,000.00
Tables	Each	280.00	5,600.00
Carts	Each	82.50	1,500.00
.12 NARCOTIC SAFES	Each	92.50	570.00
.13 REFRIGERATORS (OTHER THAN MORTUARY)			
Under Counter	Each	220.00	2,350.00
.14 TOTE TRAYS AND UTILITY CARTS	Each	77.50	460.00
.15 GAS TRACKS - Example: Operating Room	Each	450.00	3,100.00
Straight	LnFt	41.20	250.00
Curved	LnFt	51.50	300.00
.16 ANIMAL FENCING AND CAGING	SqFt	2.58	5.80
.17 DENTAL	-	-	-
.18 SURGICAL	-	-	-
.19 INCUBATORS	Each	-	4,350.00
.20 THERAPY-HEAT	Each	-	2,100.00
.21 OVERHEAD HOISTS	Each	-	2,050.00

1105.0 EDUCATIONAL EQUIPMENT

1106.0 RESIDENTIAL EQUIPMENT (NO MECH. OR ELEC.)
      See 1504 for Supply, Waste and Vent
      See 1611 for Electric Power
   .1   FREE STANDING

	UNIT	LABOR	MATERIAL
Clothes Dryer - Gas	Each	82.50	450.00
Electric	Each	82.50	400.00
Clothes Washers	Each	82.50	420.00
Compactors - 15"	Each	77.50	480.00
Disposals	Each	77.50	160.00
Dishwasher	Each	82.50	470.00
Freezers - Upright - 16 CuFt	Each	82.50	420.00
19 CuFt	Each	87.50	470.00
21 CuFt	Each	98.00	500.00
Freezers - Upright - Frost Proof			
15 CuFt	Each	82.50	530.00
18 CuFt	Each	82.50	580.00
Freezers - Chest - 15 CuFt	Each	82.50	470.00
20 CuFt	Each	92.50	580.00
25 CuFt	Each	103.00	670.00
Dehumidifier - 40 Pint	Each	82.50	175.00
Humidifier - 10 Gals/Day	Each	82.50	240.00
Range Hood	Each	72.00	170.00
Water Heater - 30 Gal. & 40 Gal.	Each	124.00	220.00
Water Softener - 30 Grains	Each	155.00	650.00

DIVISION #11 - EQUIPMENT

1106.0 RESIDENTIAL EQUIPMENT, Cont'd...

	UNIT	LABOR	MATERIAL
.1 FREE STANDING, Cont'd.			
Refrigerators			
Compact - 11 CuFt	Each	82.50	340.00
Conventional - 14 CuFt	Each	87.50	460.00
With Freezer - 15 CuFt (Also 17 & 19)	Each	92.50	570.00
21 CuFt	Each	98.00	720.00
Frostproof - 12 CuFt	Each	87.50	590.00
15 CuFt	Each	98.00	780.00
Side-by-Side - 22 CuFt	Each	98.00	1,100.00
25 CuFt	Each	103.00	1,250.00
Ranges - Electric - 30"	Each	82.50	640.00
40"	Each	87.50	1,300.00
Gas - 20"	Each	77.50	280.00
30"	Each	82.50	650.00
Microwave Ovens - Counter	Each	82.50	185.00
Over-counter	Each	124.00	430.00
.2 BUILT-IN			
Air Conditioners - 110 Volt	Each	77.50	360.00
220 Volt	Each	77.50	550.00
Dishwashers	Each	77.50	460.00
Disposals	Each	77.50	163.00
Hoods	Each	77.50	175.00
Ranges - Surface - 30" Gas	Each	77.50	280.00
42" Gas with Broiler	Each	77.50	1,250.00
30" Electric	Each	77.50	290.00
42" Elec. with Broiler	Each	82.50	1,250.00
Ovens - Single - Gas	Each	82.50	430.00
Double	Each	92.50	550.00
Single - Electric	Each	92.50	390.00
Double	Each	92.50	490.00

1107.0 ATHLETIC EQUIPMENT (M) (Carpenters)

	UNIT	LABOR	MATERIAL
.1 BASKETBALL BACKSTOPS			
Stationary	Each	360.00	660.00
Retractable	Each	520.00	1,600.00
Suspended	Each	620.00	3,100.00
Add for Electric Operated	Each	-	950.00
.2 GYM SEATING - See Div. 1306.0			
.3 SCOREBOARDS (Great Variable)	Each	980.00	5,800.00
.4 BOWLING AND BILLIARDS - See Div. 1302.0			
.5 SWIMMING POOL EQUIPMENT - See Div. 1318.0			
.6 OUTSIDE RECREATIONAL EQ. - See Div. 0208.2			
.7 GYMNASTIC EQUIPMENT - See Div. 0208.2			
.8 SHOOTING RANGE EQUIPMENT	Point	720.00	6,200.00
.9 HEALTH CLUB EQUIPMENT	-	-	-

1108.0 FOOD SERVICE EQUIPMENT (L&M) (Sheet Metal and Plumbing)

	UNIT		COST
Broilers - 36" Electric	Each	-	4,650.00
Can Washers	Each	-	3,350.00
Coffee Stand and Urn	Each	-	4,100.00
Cold Food Carts	Each	-	4,450.00
Conveyors - 12'	Each	-	5,900.00
Cutter - Mixer	Each	-	7,000.00
Dishtables - Clean	Each	-	3,600.00
Soiled	Each	-	4,850.00

DIVISION #11 - EQUIPMENT

		UNIT		COST
1108.0	**FOOD SERVICE EQUIPMENT, Cont'd...**			
	Disposer - Garbage - 3 H.P.	Each		3,800.00
	1 1/2 H.P.	Each		2,150.00
	Dishwasher- Automatic	Each		18,000.00
	Rack Type	Each		6,400.00
	Fryers - Double	Each		2,250.00
	Hot Food Carts	Each		3,500.00
	Ice Cream Dispenser	Each		1,300.50
	Ice Machine with Bin	Each		1,850.00
	Juice and Beverage Dispenser	Each		1,950.00
	Meat Saw	Each		3,300.00
	Microwave	Each		320.00
	Oven - Double Convector - Electric	Each		6,700.00
	Single	Each		3,400.00
	Portable Sink	Each		870.00
	Range - Oven- Four Burner	Each		1,400.00
	Range Hoods	Each		14,000.00
	Refrigerators	Each		2,350.00
	Silverware Dispensers	Each		220.00
	Sinks - Utility	Each		1,200.00
	Soiled Dish	Each		4,200.00
	Slicer	Each		2,350.00
	Steamers	Each		4,850.00
	Tray Dispensers	Each		690.00
	Setup Unit	Each		1,300.00
	Utility Carts	Each		210.00
	Water Stations	Each		1,050.00
	Work Tables	Each		1,100.00
	Average per Kitchen for Equipment	SqFt		62.00
1109.0	**LAUNDRY EQUIPMENT (L&M)**			
	Dryer - 50 Lb.	Each		1,950.00
	Ironer, Air	Each		6,900.00
	Hand	Each		4,750.00
	Washer Tumbler - 135 Lb.	Each		33,000.00
	50 Lb.	Each		18,000.00
1110.0	**LIBRARY EQUIPMENT (M) (Carpenters)**			
			LABOR	MATERIAL
	Book Trucks	Each	92.50	1,100.00
	Card Catalog Index - 30 drawer	Each	118.00	1,900.00
	Chairs	Each	15.50	103.00
	Charging Desks - Card File	Each	92.50	970.00
	Charging Unit	Each	92.50	1,050.00
	Open Shelf	Each	92.50	690.00
	Newspaper Rack	Each	87.50	670.00
	Shelving (Maple) - Example: 36" x 60" x 8"	LnFt	15.50	103.00
	Tables	Each	67.00	680.00
1111.0	**ECCLESIASTICAL EQUIPMENT (M) (Carpenters)**			
.1	CROSSES, SPIRES AND STEEPLES	-	-	-
.2	CONFESSIONALS - Two Person	Each	560.00	5,200.00
	One Person	Each	440.00	3,600.00
.3	PEWS (OAK) (With Kneelers)	LnFt	14.10	103.00
	Add for Frontals	LnFt	14.20	51.50
	Add for Padded Kneelers	LnFt	3.91	3.09
	Add for Birch	LnFt	-	10.30
.4	BAPTISMAL FONTS	Each	72.00	730.00
.5	MISCELLANEOUS (WOOD ONLY)			
	Main Altars	Each	430.00	5,200.00
	Side Altars	Each	240.00	1,950.00
	Lecterns	Each	144.00	1,100.50
	Credence Table	Each	113.00	760.00

DIVISION #11 - EQUIPMENT

1112.0 STAGE AND THEATRICAL EQUIPMENT (L&M) (Carpenters)

	UNIT	LABOR	MATERIAL
.1 STAGE DRAPERIES, HARDWARE, CONTROLS AND OTHER EQUIPMENT			
Motorized Rolling Curtain - 38' x 16'	Each	-	5,600.00
Stage Equipment Sets	Each	-	1,900.00
Light Bridge	Each	-	20,100.00
.2 FOLDING STAGES - Portable Thrust	Each	-	12,700.00
.3 REVOLVING STAGES - 32' Dia x 12" High	Each	-	123,600.00
.4 FIXED AUDITORIUM SEATING - Cushioned	Each	-	175.00
.5 PROJECTION SCREENS			
Motor Oper. - 8' x 8'	Each	124.00	1,450.00
10' x 10'	Each	129.00	1,700.00
Stationary - 20' x 20'	Each	155.00	2,800.00

1113.0 BANK EQUIPMENT (M) (Carpenters)

	UNIT	LABOR	MATERIAL
.1 SAFES	Each	144.00	1,050.00
.2 DRIVE-UP WINDOW UNIT - Including Cash Drawers, Counter, Window, Depository and Lock Box	Each	900.00	8,200.00
.3 TELLER UNIT	Each	300.00	2,250.00
Cash Dispensing	Each	2,850.00	38,100.00
.4 NIGHT DEPOSITORY - Envelope	Each	196.00	1,000.00
Incl. Head & Chest	Each	930.00	9,300.00
.5 SURVEILLANCE SYSTEMS			
Receival and Transmitter	Each	-	670.00
Cameras	Each	-	1,000.00
Time Lapse Recorder	Each	-	3,300.00
.6 SAFE DEPOSIT BOX	Each	7.10	72.00
Vault Doors - See Div. 0803.16			
Pneumatic Tube System - See Div. 14			

1114.0 DETENTION EQUIPMENT

1115.0 PARKING EQUIPMENT

	UNIT	LABOR	MATERIAL
.1 AUTOMATIC DEVICES			
Cash Registers	Each	530.00	10,815.00
Clocks - Rate Computing	Each	84.50	798.00
Standards	Each	69.00	638.60
Gate	Each	390.00	3,090.00
Loop Sensor	Each	210.00	952.75
Treadle	Each	160.00	381.10
Operator Station - Coin	Each	160.00	4,789.50
Key	Each	175.00	1,030.00
Radio Control - Transmitter	Each	-	1,751.00
Ticket Spitter Station	Each	220.00	5,665.00
.2 BOOTHS - Example: 4' x 6'	Each	260.00	5,562.00
.3 METERS AND POSTS	Each	82.50	309.00

1116.0 LOADING DOCK EQUIPMENT

	UNIT	LABOR	MATERIAL
.1 DOCK BUMPERS			
16" x 12" x 4"	Each	22.70	49.40
24" x 6" x 4"	Each	22.70	50.50
24" x 10" x 4"	Each	23.70	54.60
24" x 12" x 4"	Each	23.70	73.10
36" x 6" x 4"	Each	24.70	65.90
36" x 10" x 4"	Each	26.80	81.40
36" x 12" x 4"	Each	30.90	94.80
Add for Galvanized Angles	Each	-	6.40

DIVISION #11 - EQUIPMENT

		UNIT	LABOR	MATERIAL
1116.0	**LOADING DOCK EQUIPMENT, Cont'd...**			
.2	DOCK BOARDS -			
	5,000# 3' - 6" x 6' - 0"	Each	113.00	920.00
	10,000# 4' - 0" x 6' - 0"	Each	124.00	1,150.00
	15,000# 4' - 0" x 6' - 0"	Each	134.00	1,550.00
.3	LEVELERS & ADJUSTABLE RAMPS			
	8' - 0" x 8' (Add Concrete)	Each	220.00	2,800.00
.4	SHELTERS - 10' x 10' with Dock Pad	Each	300.00	1,750.00
	Strip Curtain - 10' x 10' x 8"	Each	98.00	230.00
	Dock Pad - 10' x 10'	Each	180.00	620.00
1117.0	**SERVICE STATION EQUIPMENT**			
.1	REELS - Chassis (No Mech. or Electric)	Each	124.00	620.00
	Air	Each	124.00	520.00
	Water	Each	124.00	540.00
	Motor Oil	Each	124.00	630.00
	A.T.F.	Each	124.00	680.00
	Gear Lube	Each	124.00	620.00
	Combination of Above - 3 Hose	Each	520.00	5,000.00
	Combination of Above - 5 Hose	Each	820.00	6,200.00
.2	PUMPS - Chassis	Each	144.20	550.00
	Gear Oil	Each	144.00	610.00
	A.T.F.	Each	144.00	610.00
	Oil - Motor	Each	103.00	650.00
	Gasoline	Each	310.00	1,900.00
.3	COMPRESSORS - 5 H.P.	Each	310.00	4,200.00
	3 H.P.	Each	300.00	2,000.00
	1 1/2 H.P.	Each	300.00	1,900.00
.4	TIRE CHANGERS - AUTO	Each	400.00	2,850.00
	Exhaust Systems - See Div. 15			
	Hoisting Equipment - See Div. 14			
	Tanks/Air/Water Connections - See Div. 15			
1118.0	**MUSICAL EQUIPMENT**			
1119.0	**CHECKROOM EQUIPMENT**			
	MOTORIZED, 1-Shelf, 16 ft	LnFt	18.50	113.00
	2-Shelf, 16 ft	LnFt	25.80	230.00
1120.0	**AUDIO VISUAL EQUIPMENT**			
	PROJECTION SCREEN	Each	-	124.00
	ELECTRIC CONTROLLED	Each	-	490.00
	TELEVISION CAMERA	Each	-	960.00
	MONITOR	Each	-	700.00
	VIDEO TAPE RECORDER	Each	-	1,000.00
1121.0	**INCINERATORS**			
	SMALL - 18" x 18"	Each	230.00	2,000.00
	MEDIUM - 22" x 22"	Each	250.00	2,900.00
	LARGE	Each	350.00	4,100.00
1122.0	**PHOTO AND GRAPHIC ART EQUIPMENT**			
1123.0	**WASTE HANDLING EQUIPMENT**			
	COMPACTORS - Bag - 3 CuYd Hopper	Each	880.00	11,000.00
	Cart	Each	880.00	9,800.00

DIVISION #12 - FURNISHINGS

		PAGE
1201.0	**MANUFACTURED CASEWORK (CSI 12300)**	**12-1**
	(Including Plastic Laminate Top and Hardware)	
.1	MEDICAL	12-1
.11	Wood - Prefinished	12-1
.12	Plastic Laminate	12-1
.13	Enameled Metal	12-1
.14	Stainless Steel	12-1
.2	EDUCATIONAL	12-1
1202.0	**BLINDS, SHADES AND SHUTTERS (CSI 12505)**	**12-2**
.1	BLINDS	12-2
.2	SHADES	12-2
.3	SHUTTERS	12-2
1203.0	**DRAPERIES AND CURTAIN HARDWARE (CSI 12530)**	**12-2**
1204.0	**OPEN OFFICE FURNITURE (CSI 12610)**	**12-2**
1205.0	**FLOOR MATS AND FRAMES (CSI 12690)**	**12-2**
1206.0	**AUDITORIUM AND THEATRE SEATING (CSI 12710)**	**12-2**
1207.0	**MULTIPLE USE SEATING (CSI 12710)**	**12-2**
1208.0	**BUILT-IN FOLDING TABLES AND SEATING (CSI 12745)**	**12-2**
.1	TABLES	12-2
.2	CHAIRS	12-2

1201.0 MANUFACTURED CASEWORK (L&M) (Carpenters)
(Including Plastic Laminate Top & Hardware)

		UNIT	LABOR	MATERIAL
.1	MEDICAL CASEWORK			
.11	Metal			
	Base Cabinets - 35" High x 24" Deep	LnFt	32.20	192.40
	Upper Cabinets - 24" High x 12" Deep	LnFt	31.20	156.00
	High Wall - 84" High x 12" Deep	LnFt	32.20	161.20
.12	Plastic Laminate			
	Base Cabinets - 35" High x 24" Deep	LnFt	32.20	208.00
	Upper Cabinets - 24" High x 12" Deep	LnFt	31.20	150.80
	High Wall - 84" High x 12" Deep	LnFt	32.20	239.20
.13	Wood - Prefinished			
	Base Cabinets - 35" High x 24" Deep	LnFt	32.20	192.40
	Upper Cabinets - 24" High x 12" Deep	LnFt	31.20	145.60
	High Wall - 84" High x 12" Deep	LnFt	31.20	234.00
.14	Stainless Steel			
	Base - 35" High x 24" Deep	LnFt	33.30	239.20
	Wall - 24" High x 12" Deep	LnFt	31.20	218.40
.2	EDUCATIONAL CASEWORK			

DIVISION #12 - FURNISHINGS

		UNIT	COST
1202.0	**BLINDS, SHADES AND SHUTTERS (L&M) (Carpenters)**		
.1	BLINDS - Horizontal		
	Wood 1"	SqFt	5.60
	Aluminum 1"	SqFt	5.40
	1"	SqFt	4.32
	Steel 2"	SqFt	3.90
	Cloth 1"	SqFt	6.50
	Vertical		
	Cloth or PVC 3"	SqFt	5.40
	Aluminum 3"	SqFt	6.80
.2	SHADES - Cotton	SqFt	1.92
	Vinyl Coated	SqFt	2.18
	Lightproof	SqFt	2.81
	Slat	SqFt	2.39
	Woven Aluminum	SqFt	4.16
	Fibre Glass	SqFt	2.91
	X-ray and Dark Room	SqFt	10.20
.3	SHUTTERS		
	16" x 1 1/8" x 48"	Pair	102.00
	16" x 1 1/8" x 60"	Pair	120.00
	16" x 1 1/8" x 72"	Pair	125.00
1203.0	**DRAPERIES AND CURTAIN HARDWARE (L&M) (Carpenters)**		
	TRACK	LnFt	4.78
	DRAPERIES	SqYd	8.60
	Add for Lined	SqYd	1.92
	Add for Rods	SqYd	2.18
	Add for Motorized	SqYd	6.50
1204.0	**OPEN OFFICE FURNITURE**		

		UNIT	LABOR	MATERIAL
1205.0	**FLOOR MATS AND FRAMES (M) (Carpenters)**			
	MATS - Vinyl Link - 4' x 5' x 1/2"	Each	31.20	320.00
	4' x 7' x 1/2"	Each	31.20	450.00
	Add for Color	SqFt	-	4.16
	FRAMES - Aluminum - 4' x 5'	Each	62.50	200.00
	4' x 7'	Each	73.00	250.00

		UNIT	COST
1206.0	**AUDITORIUM AND THEATRE SEATING (L&M) (Carpenters)**		
	WOOD AND PLYWOOD	Each	140.00
	CUSHIONED	Each	172.00
	ROCKING	Each	250.00
1207.0	**MULTIPLE-USE SEATING**		
	MOVABLE - Chrome and Solid Plastic	Each	67.50
	Wood and Plywood	Each	60.50
1208.0	**BUILT-IN FOLDING TABLES AND SEATING (M) (Carpenters)**		
.1	TABLES - 3' x 8' - Tempered Hardboard Top	Each	187.00
	Vinyl Laminate Top	Each	210.00
	Folding Table & Bench-Pocket Unit	Each	1,600.00
.2	CHAIRS - Metal	Each	20.80

DIVISION #13 - SPECIAL CONSTRUCTION

		PAGE
1301.0	**AIR SUPPORTED STRUCTURE (CSI 13010)**	13-2
.1	TENNIS COURTS	13-2
.2	CONSTRUCTION DOMES, WAREHOUSES, etc.	13-2
1302.0	**BOWLING ALLEYS (CSI 11500)**	13-2
1303.0	**CHIMNEYS - SPECIAL**	13-2
.1	JOB CONSTRUCTED (Radial Brick and Reinforcing Concrete) (CSI 04550)	13-2 / 13-2
.2	PREFABRICATED (Class A Metal Refractory Lined) (CSI 15600)	13-2
1304.0	**CLEAN ROOMS (CSI 13250)**	13-2
1305.0	**FLOORS - PEDESTAL OR ACCESS (CSI 10270)**	13-2
1306.0	**GRANDSTANDS, BLEACHERS AND GYM SEATING (CSI 13125)**	13-3
.1	GRANDSTANDS and BLEACHERS	13-3
.2	GYM SEATING - FOLDING	13-3
1307.0	**GREENHOUSES (CSI 13123)**	13-3
1308.0	**INCINERATORS (CSI 13400)**	13-3
1309.0	**COLD STORAGE ROOMS (CSI 13038)**	13-3
.1	CUSTOM - JOB CONSTRUCTED	13-3
.2	PREFABRICATED	13-3
1310.0	**METAL BUILDINGS (CSI 13122)**	13-4
1311.0	**RADIATION PROTECTION & RADIO FREQUENCY SHIELDING (CSI 13090 & 13095)**	13-4
.1	ELECTROSTATIC	13-4
.2	ELECTRO-MAGNETIC (Copper and Steel - Solid and Screen)	13-4
.3	X-RAY (Nuclear and Gamma)	13-4
.4	RADIO FREQUENCY SHIELDING	13-4
1312.0	**SAUNAS (CSI 13052)**	13-4
1313.0	**SCALES (CSI 10650)**	13-5
.1	FLOOR SCALES	13-5
.2	TRUCK SCALES	13-5
.3	CRANE SCALES	13-5
1314.0	**SIGNS (Custom Exterior)**	13-5
1315.0	**DOME STRUCTURES (Including Observatories) (CSI 13132)**	13-5
1316.0	**SOUND INSULATED ROOMS (CSI 13034)**	13-5
.1	ANECHOIC ROOMS	13-5
.2	AUDIOMETRIC ROOMS	13-5
1317.0	**SOUND AND VIBRATION CONTROL SYSTEMS (CSI 13081)**	13-5
1318.0	**SWIMMING POOLS (CSI 13152)**	13-6
.1	POOLS	13-6
.2	DECKS	13-6
.3	FENCES	13-6
1319.0	**TANKS**	13-6
.1	STEEL - GROUND LEVEL	13-6
.2	STEEL - ELEVATED	13-6
1320.0	**WOOD DECKS**	13-6

DIVISION #13 - SPECIAL CONSTRUCTION

1301.0 AIR SUPPORTED STRUCTURES

		UNIT	COST
.1	TENNIS COURTS - 2 LAYER THERMAL DOMES		
	(Includes Doors, Heating, Equipment, Lights & Power)		
	1 Court - 58' x 118'	SqFt	17.70
	2 Courts - 106' x 118'	SqFt	16.70
	3 Courts - 156' x 118'	SqFt	15.10
	4 Courts - 201' x 118'	SqFt	14.00
	Deduct for Non Thermal Type	SqFt	2.08
.2	CONSTRUCTION DOMES, WAREHOUSES, ETC.	SqFt	7.80
	Add for Heating	SqFt	3.12
	Add for Light and Power	SqFt	1.56
	Add for Air Lock Doors	Each	13,800.00
	Add for Revolving Doors	Each	7,100.00

1302.0 BOWLING ALLEYS (L&M) (Carpenters) — Lane — 38,000.00
    Add for Automatic Scorers — Lane — 15,100.00

1303.0 CHIMNEYS - SPECIAL (Foundations Not Included)

.1 JOB CONSTRUCTED (L&M) (Bricklayers)

Inside Diam.	Height		Unit	Cost
5'	100'	Radial Brick	Each	124,800.00
6'	150'	Radial Brick	Each	150,800.00
7'	200'	Radial Brick	Each	197,600.00
8'	250'	Reinforced Concrete	Each	478,400.00
9'	200'	Reinforced Concrete	Each	592,800.00
10'	250'	Reinforced Concrete	Each	811,200.00
12'	300'	Reinforced Concrete	Each	1,352,000.00

.2 PREFABRICATED (METAL REFRACTORY LINED)
1800° - 2000° (L&M) (Sheet Metal & Iron Workers)

	Unit	Cost
10"	LnFt	56.00
12"	LnFt	60.50
15"	LnFt	72.00
18"	LnFt	87.50
21"	LnFt	104.00
24"	LnFt	125.00
30"	LnFt	220.00
36"	LnFt	310.00
Add for Cleanouts or T's	Each	320.00
Add for V.L. Label	Each	10%
Deduct for 800° Chimney	Each	5%

1304.0 CLEAN ROOMS — SqFt — 200.00

1305.0 FLOORS - ACCESS OR PEDESTAL (L&M) (Carpenters)
(Includes Laminate Topping)

	Unit	Cost
Aluminum	SqFt	19.80
Steel - Galvanized	SqFt	12.50
Plywood - Metal Covered	SqFt	11.50
Add for Carpeting	SqFt	3.12
Add for Stairs and Ramps	SqFt	50%

DIVISION #13 - SPECIAL CONSTRUCTION

		UNIT	COST
1306.0	**GRANDSTANDS, BLEACHERS AND GYM SEATING**		
.1	GRANDSTANDS AND BLEACHERS		
	Not Permanent	Each Seat	24.40
	Permanent - Aluminum	Each Seat	98.00
	Steel	Each Seat	62.50
	Concrete	Each Seat	230.00
.2	GYM SEATING - FOLDING	Each Seat	68.50
1307.0	**GREENHOUSE (L&M) (Glaziers)**		
	SINGLE ROOM TYPE - Commercial	SqFt	21.80
	Educational	SqFt	29.10
	Residential	SqFt	32.20
	Add for Heating, Lighting and Benches	SqFt	17.70
	Add for Cooling	SqFt	7.50
1308.0	**INCINERATORS (L&M) (Ironworkers and Bricklayers)**		
	Stack and Breaching Not Included		
	Pounds/Hour		
	100	Each	8,400.00
	200	Each	8,500.00
	300	Each	9,500.00
	400	Each	10,700.00
	500	Each	11,800.00
	600	Each	15,100.00
	800	Each	24,400.00
	Add for Gas Burner	Each	1,600.00
	Add for Controls	Each	880.00
	Add for Piping	Each	900.00
1309.0	**INSULATED ROOMS (L&M) (Carpenters)**		
.1	CUSTOM - Cooler	SqFt	172.00
	Freezer	SqFt	198.00
.2	PREFABRICATED (SELF-CONTAINED)		
	Cooler Example: Room 7'6" x 9'6"		
	Wood (Plywood)	Each	6,200.00
	Metal Covered 20 ga Galvanized #3 SS	Each	7,700.00
	Add for 3/4 HP Compressor	Each	1,050.00
	Add for 3/4 HP Compressor Starter & Coils	Each	4,750.00
	Add for Electrical	Each	810.00
	Cooler - Total Cost - Wood	SqFt	172.00
	Metal Covered	SqFt	192.00
	Freezer Example: Room 7'6" x 9'6"		
	Wood (Plywood)	Each	6,400.00
	Metal Covered 20 ga Galvanized #3 SS	Each	8,000.00
	Add for 3/4 HP Compressor	Each	1,400.00
	Add for 3/4 HP Compressor Starter & Coils	Each	4,300.00
	Add for Electrical	Each	830.00
	Freezer - Total Cost - Wood	SqFt	187.00
	Metal Covered	SqFt	220.00
	Add for Shelving	SqFt	8.80

DIVISION #13 - SPECIAL CONSTRUCTION

		UNIT	COST
1310.0	**METAL BUILDINGS**		
.1	SHELL - FLOOR AREA (NO FOUNDATION)	SqFt	14.00
	24 ga. 12' Sidewalls and 60' Span (Galvanized)		
	Add per foot above 12' - Floor Area	SqFt	.40
	Add per foot of Span above or below 60'	SqFt	.24
	Add for 2' Overhand - Overhang Area	SqFt	5.40
	Add for Aluminum	SqFt	.68
	Add for Liner Panels - Wall Area	SqFt	2.03
	Add for Insulation 3 1/2" - Wall Area	SqFt	.78
	Add for Insulation 6" - Ceiling Area	SqFt	.99
	Add for Door Openings - Including Hardware	Each	700.00
	Add for Window Openings	Each	370.00
	Add for Insulated Glass Area	SqFt	10.40
	Add for Overhead Door Area	SqFt	9.60
	See 0803.4 for Door Cost		
	Add for Gutter	LnFt	4.68
	FOUNDATION & GROUND SLAB COST AVG (3-Foot Frost)	SqFt	8.10
1311.0	**RADIATION PROTECTION (L&M) (Carpenters)**		
.1	ELECTROSTATIC SHIELDING		
	Custom - Vinyl Sandwiched Copper Mesh for Floors, Walls & Ceilings (including Copper Foil Tabs & Plastic Tape)	SqFt	23.90
	Add per door and Frame with RF Shielded Glass (including Grounding Wires)	Each	3,650.00
	Add per Borrowed Light RF Shielded Glass	Each	1,450.00
	Example: Room 10' x 8' (Prefab)	Room	28,100.00
		or SqFt	360.00
.2	ELECTROMAGNETIC SHIELDING		
	Example: Room 10' x 8' (Prefab)	Room	12,000.00
		or SqFt	156.00
	Add for Finish, Electricity and Erection	Room	9,880.00
.3	X-RAY		
	Deep Therapy		
	Example: Room 10' x 15'	Room	12,500.00
		or SqFt	83.00
	Radiography or Fluoroscopy		
	Example: Room 10' x 15'	Room	5,300.00
		or SqFt	35.40
.4	RADIO FREQUENCY SHIELDING		
	Example: Room 10' x 10'	Room	31,200.00
		or SqFt	3,100.00
	See 1104.10 for X-Ray Equipment		
1312.0	**SAUNAS (M) (Carpenters)**		
	Examples (includes Heaters):		
	Room 5' x 6' Cedar (5 K.W. Unit)	Room	4,700.00
		or SqFt	156.00
	Room 6' x 6' Cedar (6 K.W. Unit)	Room	5,600.00
		or SqFt	156.00
	Room 6' x 7' Cedar (7.5 K.W. Unit)	Room	5,600.00
		or SqFt	135.00
	Add for Electrical Connections and Ventilating		780.00

DIVISION #13 - SPECIAL CONSTRUCTION

1313.0 SCALES (NO PORTABLE) (Ironworkers)

.1 FLOOR SCALES (HEAVY DUTY INDUSTRIAL)

Size	Capacity	Labor	Material	Concrete Work
46" x 38"	800 lb.	430.00	4,250.00	-
48" x 48"	1,350 lb.	460.00	4,500.00	-
60" x 48"	1,600 lb.	640.00	4,800.00	-
72" x 48"	2,600 lb.	660.00	5,600.00	-
76" x 54"	6,000 lb.	1,150.00	5,900.00	-
84" x 60"	10,000 lb.	1,900.00	7,800.00	-

.2 TRUCK SCALES - FULL WITH WEIGH BEAM

Truck (4-Section) -

Size	Capacity	Labor	Material	Concrete Work
45" x 10"	50 Ton	2,700.00	21,200.00	20,800.00
50" x 10"	50 Ton	3,000.00	21,800.00	21,300.00
60" x 10"	50 Ton	3,450.00	22,900.00	21,800.00
70" x 10"	50 Ton	4,050.00	23,900.00	23,300.00
Add for 60-Ton Capacity			3,450.00	

Axle Load -

Size	Capacity	Labor	Material	Concrete Work
8" x 10"	20 Ton	1,900.00	9,700.00	7,000.00
8" x 10"	30 Ton	2,450.00	10,600.00	7,100.00
10" x 10"	20 Ton	2,600.00	11,100.00	7,600.00
10" x 10"	30 Ton	3,000.00	12,400.00	7,700.00
Add to Above for Printer and Dial			2,700.00	

.3 CRANE SCALES -

	Capacity	Labor	Material	Concrete Work
	1 Ton	146.00	2,500.00	-
	5 Ton	280.00	2,700.00	-

1314.0 SIGNS (Special Exterior) (L&M) (Sheet Metal Workers)

1315.0 SKYDOMES (L&M) (Carpenters)

EXTRUDED ALUMINUM FRAME, WIRE GLASS
EXTERIOR AND FIBERGLASS SUB CEILING

	UNIT	COST
12' x 18' Ridge Type	Each	8,200.00
	or SqFt	38.50
18' x 18' Pyramid Type	Each	15,800.00
	or SqFt	48.90
36' Diameter Geometric Domes	Each	64,500.00
	or SqFt	62.50
18' Diameter Revolving Domes	Each	16,600.00
	or SqFt	60.50

1316.0 SOUND INSULATED ROOMS (Including Vibration Control) (M) (Carpenters)

.1 ANECHOIC ROOMS (ISOLATORS, PANELS AND WEDGES)

	Room	121,700.00
Example: Room 30' x 30'	or SqFt	135.00

.2 AUDIOMETRIC ROOM (PREFABRICATED PANELS)

Example: Research Room - 5' x 8'	Room	10,800.00
	or SqFt	270.00
Example: Medical Practice Room - 5' x 8'	Room	20,000.00
	or SqFt	500.00

1317.0 SOUND AND VIBRATION CONTROL SYSTEMS

DIVISION #13 - SPECIAL CONSTRUCTION

1318.0 SWIMMING POOLS

		UNIT	COST
.1	POOLS		
	<u>Deep Pools</u> - 3 to 9 feet		
	(Incl. Filters, Skimmers, Lights, Clng. Equip.)		
	Rectangular or L Shape:		
	Residential - 1,000 SqFt	Each	38,500.00
		or SqFt	38.50
	Motels and Apartments - 2,000 SqFt	Each	89,400.00
		or SqFt	44.70
	Municipal - 5,000 SqFt	Each	260,000.00
		or SqFt	52.00
	Add for Diving Stand - Steel - 1 Meter	Each	3,100.00
	Steel - 3 Meter	Each	4,600.00
	Add for Diving Boards - 12' Fiberglass	Each	850.00
	16' Aluminum	Each	1,550.00
	Add for Slides - 6' Fiberglass	Each	810.00
	12' Fiberglass	Each	1,250.00
	Add for Ladders - 10'	Each	310.00
	Add for Lifeguard Chairs	Each	1,050.00
	Add for Covers	Each	760.00
	<u>Shallow Pools</u> - 3 to 5 feet -		
	Gunite with Plaster Finish		
	Residential - 1,000 SqFt	Each	29,100.00
		or SqFt	29.10
	Motels and Apartments - 2,000 SqFt	Each	74,900.00
		or SqFt	37.40
	Add for Water, Drainage & Power to Pools	Each	-
	Deduct for Vinyl Lined Pools	SqFt	9.40
	Add for Free Form Pools	Each	5%
	Add for Heaters	Each	3,350.00
	Add for Automatic Cleaners	Each	3,100.00
	<u>Wading Pools</u> - 12' x 12' or 9' Round	Each	4,700.00
		or SqFt	33.30
.2	DECKS - CONCRETE - Average 10' around Pool	Each	3,750.00
	and Average 1,000 SqFt	or SqFt	3.74
	Add for Tile	SqFt	7.50
	Add for Brick	SqFt	6.70
.3	FENCES - 4' ALUMINUM OR GALVANIZED	LnFt	1,050.00

1319.0 TANKS

		UNIT	COST
.1	STEEL - GROUND LEVEL (NO FOUNDATION)		
	150,000 Gals	Each	133,100.00
	500,000 Gals	Each	223,600.00
	1,000,000 Gals	Each	369,200.00
	2,000,000 Gals	Each	582,400.00
.2	STEEL - ELEVATED		
	100,000 Gals	Each	286,000.00
	250,000 Gals	Each	384,800.00
	500,000 Gals	Each	624,000.00
	1,000,000 Gals	Each	1,352,000.00

1320.0 WOOD DECKS - WITH RAILS

	UNIT	COST
CEDAR	SqFt	14.30
TREATED FIR	SqFt	12.80
Add for Seats	LnFt	3.64
Add for Stairs	Riser	47.80

DIVISION #14 - CONVEYING SYSTEMS

			PAGE
1401.0	**ELEVATORS (Passenger, Freight & Service) (CSI 14200)**		14-2
.1	ELECTRIC (CABLE OR TRACTION)		14-2
.11	Passenger		14-2
.12	Freight		14-2
.2	OIL HYDRAULIC		14-2
.21	Passenger		14-2
.22	Freight		14-2
.3	RESIDENTIAL (WINDING DRUM) (CSI 14235)		14-2
1402.0	**MOVING STAIRS AND WALKS**		14-3
.1	MOVING STAIRS (ESCALATORS) (CSI 14200)		14-3
.2	MOVING WALKS (CSI 14320)		14-3
1403.0	**DUMBWAITERS (CSI 14100)**		14-3
.1	ELECTRIC		14-3
.2	HAND OPERATED		14-3
1404.0	**LIFTS (CSI 14400)**		14-3
.1	WHEELCHAIR LIFTS		14-3
.2	STAIR LIFTS		14-3
.3	VEHICLE - HYDRAULIC LIFTS		14-3
.4	MATERIAL HANDLING LIFTS		14-3
.5	STAGE LIFTS		14-3
1405.0	**PNEUMATIC TUBE SYSTEMS (CSI 14580)**		14-3
1406.0	**MATERIAL HANDLING (CSI 14500)**		14-3
1407.0	**TURNTABLES (CSI 14700)**		14-3
1408.0	**POWERED SCAFFOLDING (CSI 14840)**		14-3

DIVISION #14 - CONVEYING SYSTEMS

1401.0 ELEVATORS (L&M) (Elevator Constructors & Ironworkers)

.1   ELECTRIC (CABLE/ TRACTION)

.11   Passenger - Std Specifications

	Speed to F.P.M.	Capacity and Cost 2,000#	3,000#	4,000#	10,000#
Geared A.C. Rheostatic:					
Selective Collective Operation					
3 Openings	125	66,000	-	-	-
4 Openings	125	69,000	-	-	-
5 Openings	125	72,000	-	-	-
Add per Floor	-	5,000	-	-	-
Geared Variable Voltage:					
Selective Collective Operation					
8 Openings	250	121,000	124,000	127,000	-
	350	125,000	129,000	130,000	-
10 Openings	250	124,000	128,000	132,000	-
	350	128,000	126,000	135,000	-
Add per Floor	-	6,500	7,000	7,500	-
Group Supervisory					
8 Openings	250	-	134,000	136,000	-
	350	-	142,000	141,000	-
10 Openings	250	-	134,000	139,000	-
	350	-	141,000	142,000	-
Add per Floor	-	-	7,000	7,500	-
Gearless Variable Voltage:					
Group Supervisory					
8 Openings	500	-	167,000	163,000	-
	800	-	180,000	176,000	-
	1000	-	190,000	190,000	-
10 Openings	500	-	170,000	152,000	-
	800	-	185,000	189,000	-
	1000	-	200,000	190,000	-
Add per Floor	-	-	6,800	7,000	-
Add per Fl. of Express Zone	-	6,300	6,700	-	-

.12   Freight

	Speed	2,000#	3,000#	4,000#	10,000#
3 Openings	100	-	-	88,000	93,000
4 Openings	100	-	-	93,000	98,000
Add per Floor	-	-	-	19,800	11,600
Add per Power Operated Door	-	-	-	-	7,000

.2   OIL HYDRAULIC

.21   Passenger

	Speed	2,000#	3,000#	4,000#	10,000#
2 Openings	100	47,000	-	-	-
3 Openings	100	51,000	-	-	-
4 Openings	100	58,000	-	-	-
5 Openings	100	65,000	-	-	-
6 Openings	100	74,000	-	-	-

Deduct for Limited Use Elevator, non-Commercial 8,000# - $5,000

.22   Freight

	Speed	2,000#	3,000#	4,000#	10,000#
2 Openings	100	-	-	55,000	67,000
3 Openings	100	-	-	64,000	71,000
4 Openings	100	-	-	66,000	75,000

Add per Power Operated Door - $11,000

.3   RESIDENTIAL (WINDING DRUM)

     2 Openings - Speed to 36 F.P.M., 700# Capacity - $19,000
         Speed to 30 F.P.M., 410# Capacity - $14,500
     Add for 3 Openings - $3,300

Add for ADA Updating - $12,000
Add for Updating Generators, Controllers & Operating Equipment - $55,000
Add to 1401.1 and 1401.2 Above for Custom Specifications - 10%
Add to 1401.2 for Shaft Work, Electric Power & Mechanical - approx. $53,000
Add for Inspections and Electric Work - $1,400

DIVISION #14 - CONVEYING SYSTEMS

1402.0 MOVING STAIRS & WALKS (L&M) (Elevator Constructors & Ironworkers)

   .1 MOVING STAIRS (Escalators)

	UNIT	CAPACITY	COST
16' Floor Height			
Metal Balustrade 32" Wide			115,400
48"			126,900
Glass Balustrade 32" Wide			125,800
48"			137,300
Add or Deduct per Foot Floor Height			2,800
Add for Over 20 Feet Floor Height			25%

   .2 MOVING WALKS

	UNIT	CAPACITY	COST
Belt Type 32"	LnFt		750
48"	LnFt		820
Pallet Type 32"	LnFt		810
48"	LnFt		880

1403.0 DUMBWAITERS (L&M) (Elevator Constructors)

   .1 ELECTRIC

Drum Type	Speed F.P.M.		
2 Openings - 1 Stop	25 to 75	50# to 500#	13,500
3 Openings - 2 Stops	25 to 75	50# to 500#	16,600
Add per Floor			3,650
Traction Type			
3 Openings - 2 Stops	50 to 300	50# to 500#	18,200
4 Openings - 3 Stops	50 to 300	50# to 500#	20,600
5 Openings - 4 Stops	50 to 300	50# to 500#	25,000
6 Openings - 5 Stops	50 to 300	50# to 500#	27,600
Add per Floor			3,650

   .2 HAND OPERATED

	COST
2 Openings - 1 Stop	6,200
3 Openings - 2 Stops	6,900
Add per Floor	940
Add for all Stainless Steel	25%

1404.0 LIFTS (L&M) (Ironworkers & Elevator Constructors)

	UNIT	CAPACITY	COST
.1 WHEELCHAIR LIFTS - 42"	Each	1,050#	17,200
96"	Each	1,400#	20,800
.2 STAIR LIFTS (CLIMBERS)	Each		5,600
.3 VEHICLE - SEMI HYDRAULIC LIFTS - 1 Post	Each	8,000#	7,100
2 Post	Each	10,000#	11,200
2 Post	Each	15,000#	17,200
Add for Fully Hydraulic - 2 Post	Each	10,000#	3,950
.4 MATERIAL HANDLING LIFTS	Each		-
.5 STAGE LIFTS - 12' x 30'	Each		85,300
Add for Permits, Inspections & Elec Hookups			1,450

See 1116 for Loading Dock Lift Equipment

1405.0 PNEUMATIC TUBE SYSTEMS (L&M) (Sheet Metal Workers)

	UNIT	COST
3" ROUND - Single Tube	LnFt	28
Twin Tube	LnFt	37
4" ROUND - Single Tube	LnFt	29
Twin Tube	LnFt	38
4" x 7" OVAL - Single Tube	LnFt	36
Twin Tube	LnFt	50

1406.0 MATERIAL HANDLING (L&M) (Ironworkers & Millwrights)   -

1407.0 TURNTABLES (L&M) (Ironworkers & Millwrights)   -

1408.0 POWERED SCAFFOLDING (L&M) (Ironworkers)   -

DIVISION #14 - CONVEYING SYSTEMS

14-4

DIVISION #15 - MECHANICAL

Wage Rates (Including Fringes) & Location Modifiers
July 2002-2003

	Metropolitan Area		Plumber Wage Rate	Wage Rate Location Modifier
1.	Akron	*	34.30	106
2.	Albany-Schenectady-Troy	*	33.50	100
3.	Atlanta		30.40	88
4.	Austin	**	25.30	74
5.	Baltimore		34.30	101
6.	Birmingham		25.20	70
7.	Boston		44.20	135
8.	Buffalo-Niagara Falls	*	33.50	102
9.	Charlotte	**	21.40	60
10.	Chicago-Gary		42.70	122
11.	Cincinnati		32.40	98
12.	Cleveland		40.00	120
13.	Columbus	*	35.80	95
14.	Dallas-Fort Worth	**	25.10	75
15.	Dayton	*	32.20	97
16.	Denver-Boulder		29.00	90
17.	Detroit		43.60	125
18.	Flint	*	34.40	102
19.	Grand Rapids	**	32.70	96
20.	Greensboro-West Salem	**	21.30	60
21.	Hartford-New Britain	*	37.70	115
22.	Houston	**	28.00	78
23.	Indianapolis	*	36.20	106
24.	Jacksonville	**	27.40	84
25.	Kansas City		37.90	100
26.	Los Angeles-Long Beach	*	40.30	131
27.	Louisville	**	29.40	87
28.	Memphis	**	26.50	91
29.	Miami	**	29.70	87
30.	Milwaukee	*	36.40	110
31.	Minneapolis-St. Paul		37.80	109
32.	Nashville	**	27.20	76
33.	New Orleans	**	24.40	68
34.	New York		62.00	197
35.	Norfolk-Portsmouth	**	26.90	79
36.	Oklahoma City	**	27.90	82
37.	Omaha-Council Bluffs	*	29.20	86
38.	Orlando	**	28.60	85
39.	Philadelphia	*	43.90	139
40.	Phoenix	**	31.50	95
41.	Pittsburgh	*	37.00	111
42.	Portland	*	39.20	106
43.	Providence-Pawtucket	*	35.00	104
44.	Richmond	**	26.80	78
45.	Rochester	**	36.00	108
46.	Sacramento	*	51.50	164
47.	St. Louis		41.00	105
48.	Salt Lake City-Ogden	**	26.78	78
49.	San Antonio	**	28.40	81
50.	San Diego	*	40.20	120
51.	San Francisco-Oakland-San Jose	*	51.50	164
52.	Seattle-Everett		43.90	133
53.	Springfield-Holyoke-Chicopee	*	34.50	103
54.	Syracuse	*	34.20	97
55.	Tampa-St. Petersburg	**	29.80	85
56.	Toledo	*	36.70	110
57.	Tulsa	**	28.70	82
58.	Tucson	**	32.40	96
59.	Washington D.C.	*	34.60	88
60.	Youngstown-Warren	*	32.90	97
	AVERAGE		**34.02**	

Impact Ratio: Labor 35%, Material 65%
* Contract Not Settled - Wage Interpolated or Unknown
** Non Signatory or Open Shop Rate

DIVISION #15 - MECHANICAL

		PAGE
1501.0	**BASE MATERIALS AND METHODS (CSI 15050)**	**15-4**
.1	PIPE AND PIPE FITTINGS	15-4
.2	PIPING SPECIALITIES	15-6
.21	Expansion Joints	15-6
.22	Vacuum Breakers	15-6
.23	Strainers	15-6
.3	MECHANICAL - SUPPORTING ANCHORS AND SEALS	15-6
.4	VALVES	15-6
.5	PUMPS	15-7
.6	VIBRATION ISOLATION	15-7
.7	METERS AND GAGES	15-7
.8	TANKS	15-7
1502.0	**MECHANICAL INSULATION**	**15-7**
.1	INSULATION FOR PIPING	15-7
.2	INSULATION FOR DUCT WORK	15-7
.3	INSULATION FOR BOILERS	15-7
1503.0	**FIRE PROTECTION EQUIPMENT (CSI 15300)**	**15-8**
.1	AUTOMATIC SPRINKLER EQUIPMENT	15-8
.2	CARBON DIOXIDE EQUIPMENT	15-8
.3	STANDPIPE AND FIRE HOSE EQUIPMENT	15-8
1504.0	**PLUMBING (CSI 15400)**	**15-8**
.1	EQUIPMENT	15-8
.2	PACKAGE WASTE, VENT, OR WATER PIPING	15-8
.3	DOMESTIC WATER SOFTENER	15-8
.4	PLUMBING FIXTURES	15-8
.5	RESIDENTIAL PLUMBING	15-8
.6	POOL EQUIPMENT	15-8
.7	FOUNTAIN PIPING	15-8
1505.0	**HEAT GENERATION (CSI 15500)**	**15-9**
.1	FUEL HANDLING EQUIPMENT	15-9
.2	ASH REMOVAL SYSTEM	15-9
.3	LINED BREECHINGS	15-9
.4	BOILERS	15-9
1506.0	**REFRIGERATION (CSI 15600)**	**15-9**
.1	REFRIGERANT COMPRESSORS	15-9
.2	CONDENSERS	15-9
.3	CHILLERS	15-9
.4	COOLING TOWERS	15-9
1507.0	**HEAT TRANSFERS (CSI 15750)**	**15-9**
.1	HOT WATER SPECIALITIES	15-9
.2	CONDENSATE PUMPS AND RECEIVER SETS	15-9
.3	HEAT EXCHANGERS	15-9
.4	TERMINAL UNITS	15-9
.5	COILS	15-10
.6	UNIT HEATERS	15-10
.7	PACKAGED HEATING AND COOLING	15-10
.8	STEAM SPECIALTIES	15-10

DIVISION #15 - MECHANICAL

		PAGE

1508.0 AIR HANDLING AND DISTRIBUTION (CSI 15850 AND 15880) — 15-10

    .1   FURNACE — 15-10
    .2   FANS — 15-10
    .3   DUCT WORK — 15-10
    .4   DUCT ACCESSORIES — 15-10
    .5   RESIDENTIAL HEATING AND AIR CONDITIONING — 15-10

1509.0 SEWAGE DRAINAGE (CSI 02700) — 15-10

    .1   SEWAGE EJECTORS — 15-10
    .2   GREASE INTERCEPTORS — 15-10
    .3   LIFT STATIONS - STEEL, CONCRETE OR FIBERGLASS — 15-10
    .4   SEPTIC TANKS AND DRAIN FIELD — 15-10
    .5   SEWAGE TREATMENT EQUIPMENT - STEEL OR CONCRETE — 15-10
    .6   AERATION EQUIPMENT - STEEL — 15-10
    .7   SLUDGE DIGESTION — 15-10

1510.0 CONTROLS AND INSTRUMENTATION (CSI 15950) — 15-10

    .1   ELECTRICAL AND INTERLOCKS — 15-10
    .2   IDENTIFICATION — 15-10
    .3   CONTROL PIPING, TUBING AND WIRING — 15-10
    .4   CONTROL AIR COMPRESSOR AND DRYER — 15-10
    .5   CONTROL PANELS — 15-10
    .6   INSTRUMENT PANELBOARD — 15-10
    .7   PRIMARY CONTROL DEVICES — 15-10
    .8   RECORDING DEVICES — 15-10
    .9   ALARM DEVICES — 15-10

1511.0 TESTING, ADJUSTING AND BALANCING (CSI 15990) — 15-10

DIVISION #15 - MECHANICAL

Costs of mechanical work are priced as total contractor's costs with an overhead and fee of 20% included. Not included are scaffold, hoisting, temporary heat, enclosures and bonding. Included are 35% taxes and insurance on labor, equipment, tools and 5% sales tax.

1501.0 BASE MATERIALS AND METHODS

	UNIT	COST
.1 PIPE AND PIPE FITTINGS (Including Fittings)		
Steel - Black - Threaded 3/4"	LnFt	2.99
1"	LnFt	3.47
1-1/2"	LnFt	4.94
2"	LnFt	9.50
2-1/2"	LnFt	13.10
3"	LnFt	18.20
4"	LnFt	22.80
5"	LnFt	26.30
6"	LnFt	34.10
8"	LnFt	49.90
Steel - Galvanized Schedule 40 3/4"	LnFt	6.30
1"	LnFt	8.00
1-1/4"	LnFt	9.20
1-1/2"	LnFt	10.90
2"	LnFt	14.20
3"	LnFt	22.80
4"	LnFt	30.70
5"	LnFt	37.10
6"	LnFt	50.50
8"	LnFt	56.50
Deduct for Black Iron	LnFt	5%
Deduct for Underground	LnFt	10%
Cast Iron - Hub and Heavy Duty 2"	LnFt	14.50
3"	LnFt	15.20
4"	LnFt	21.00
5"	LnFt	25.00
6"	LnFt	34.20
8"	LnFt	40.40
Add for Line Pipe	LnFt	4.46
Deduct for Hubless Type	LnFt	10%
Deduct for Standard	LnFt	15%
P.V.C. - Schedule 40 1/2"	LnFt	4.46
3/4"	LnFt	4.94
1"	LnFt	5.50
1-1/4"	LnFt	5.80
1-1/2"	LnFt	6.50
2"	LnFt	7.10
3"	LnFt	8.20
4"	LnFt	14.00
6"	LnFt	24.20
Deduct for Underground	LnFt	10%
P.V.C. - Type L 3/8"	LnFt	3.31
1/2"	LnFt	4.41
3/4"	LnFt	5.00
1"	LnFt	5.90
1-1/4"	LnFt	6.50
1-1/2"	LnFt	7.40
2"	LnFt	8.20
3"	LnFt	9.50
4"	LnFt	14.20
6"	LnFt	27.30
Add for Type K (Underground)	LnFt	20%
Deduct for Type M (Residential)	LnFt	5%

DIVISION #15 - MECHANICAL

1501.0 BASE MATERIALS AND METHODS (Cont'd...)

.1   PIPE AND PIPE FITTINGS (Cont'd...)   <u>UNIT</u>   <u>COST</u>

	Unit	Cost
Brass - Threadless 1/2"	LnFt	7.80
3/4"	LnFt	8.60
1"	LnFt	10.80
1-1/4"	LnFt	13.70
1-1/2"	LnFt	16.30
2"	LnFt	19.70
3"	LnFt	27.30
4"	LnFt	35.70
Add for Threaded	LnFt	30%
Stainless Steel Schedule 40 1/4"	LnFt	7.50
1/2"	LnFt	10.70
3/4"	LnFt	14.10
1"	LnFt	16.50
2"	LnFt	22.80
4"	LnFt	56.00
Add for Type 316	LnFt	20%
Fibre Glass - 1"	LnFt	15.50
1-1/2"	LnFt	17.70
2"	LnFt	20.10
3"	LnFt	23.10
4"	LnFt	26.60
5"	LnFt	34.00
6"	LnFt	51.50
8"	LnFt	59.50

Utility Piping -

	UNIT	Reinforced Concrete	Vitrified Clay	Bituminous Coated Corrugated	Smooth Paved Corrugated	Ductile Iron
6"	LnFt	$6.20	$6.50	$8.70	$9.50	$14.70
8"	LnFt	6.70	9.10	9.20	10.00	21.30
1"	LnFt	8.50	11.90	10.80	12.00	24.50
12"	LnFt	10.40	14.10	13.10	14.70	34.20
15"	LnFt	12.50	20.60	16.30	17.40	44.30
18"	LnFt	16.10	30.70	20.30	30.10	57.50
21"	LnFt	20.30	37.40	23.80	26.30	69.00
24"	LnFt	25.50	54.50	26.30	31.10	81.50
27"	LnFt	34.30	76.50	31.80	37.00	-
30"	LnFt	43.10	100.90	33.60	42.50	-
36"	LnFt	56.50	138.00	37.10	52.00	-
42"	LnFt	74.50	-	43.10	60.00	-
48"	LnFt	93.50	-	59.00	67.00	-
54"	LnFt	109.00	-	76.00	89.50	-
60"	LnFt	130.00	-	94.50	113.00	-

Add for Trenching and Backfill - See Division 0202.0
See 0205 for Site Drainage Items

DIVISION #15 - MECHANICAL

1501.0 BASE MATERIALS AND METHODS, Cont'd...

.2   PIPING SPECIALTIES

.21   Expansion Joints - Copper

	UNIT	COST
1/2"	Each	$54.50
3/4"	Each	82.00
1"	Each	130.00
1-1/2"	Each	270.00
2"	Each	410.00
3"	Each	760.00
4"	Each	830.00

.22   Vacuum Breakers

	UNIT	COST
1/2"	Each	34.70
3/4"	Each	40.40
1"	Each	54.50
1-1/2"	Each	75.00

.23   Strainers - Screwed Ends

	UNIT	IRON BODY	BRONZE BODY
1/2"	Each	29.40	37.80
3/4"	Each	33.60	43.60
1"	Each	40.40	59.00
1-1/2"	Each	55.50	93.50
2"	Each	89.50	152.00
3"	Each	152.00	294.00
4"	Each	410.00	10,800.00

Strainers - Flanged Ends

	UNIT	IRON BODY	BRONZE BODY
2"	Each	142.00	390.00
3"	Each	250.00	800.00
4"	Each	400.00	1,400.00
5"	Each	640.00	1,900.00
6"	Each	830.00	2,650.00

.3   MECHANICAL - SUPPORTING ANCHORS & SEALS

.4   VALVES

Bronze Body	UNIT	150 psi Gate	150 psi Check	150 psi Globe	Relief	300 psi to 75 Reducing
1/2"	Each	$27.00	$23.90	$31.20	$45.80	$64.00
3/4"	Each	33.30	27.00	41.60	55.00	76.00
1"	Each	39.50	31.20	55.00	81.00	112.00
1-1/2"	Each	62.50	51.00	87.00	260.00	250.00
2"	Each	81.00	77.00	125.00	290.00	370.00
2-1/2"	Each	144.00	440.00	210.00	-	-
3"	Each	198.00	167.00	320.00	-	-

Iron Body	UNIT	125 psi Gate	125 psi Check	125 psi Globe
2"	Each	$230.00	$182.00	$330.00
2-1/2"	Each	240.00	185.00	380.00
3"	Each	270.00	189.00	450.00
4"	Each	400.00	340.00	600.00
6"	Each	570.00	360.00	760.00

Steel	UNIT	Gate	Check	Globe
2"	Each	$780.00	$750.00	$970.00
2-1/2"	Each	1,206.40	1,000.00	1,400.00
3"	Each	1,216.80	1,000.00	1,500.00
4"	Each	1,456.00	1,400.00	2,100.00

Plastic	UNIT	Ball	Foot
1/2"	Each	$22.40	$52.50
3/4"	Each	27.00	58.00
1"	Each	31.20	76.00
1-1/2"	Each	53.00	123.00
2"	Each	60.30	146.00

DIVISION #15 - MECHANICAL

1501.0 BASE MATERIALS AND METHODS, Cont'd... UNIT   COST

.5   PUMPS

	Unit	Cost
Condensate Return - Simplex	Each	$ 1,700.00
Duplex	Each	4,500.00
Utility Water Pump -		
Multi - Two Stage 3" & 4" 75 HP	Each	17,200.00
Four Stage 3" & 4" 150 HP	Each	31,700.00
Single - End Suction 1" x 2" 3 HP	Each	4,400.00
End Suction 2" x 3" 15 HP	Each	5,600.00
Condenser Water Pump, Pressure - 100 GPM	Each	15,100.00
300 GPM	Each	20,000.00
1200 GPM	Each	43,900.00
2000 GPM	Each	54,100.00
Submersible Pump 2"	Each	620.00
Pedestal Pump	Each	192.00

.6   VIBRATION ISOLATION 5" long   Each   320.00
        6" long   Each   610.00

.7   METERS AND GAGES

	Unit	Cost
Disc Type - 3/4"	Each	156.00
1"	Each	192.00
2"	Each	530.00
Compound Type - 3"	Each	1,850.00
4"	Each	2,900.00

.8   TANKS

	Unit	Cost
Fuel Storage - Steel - 1,000 Gallons	Each	1,800.00
2,000	Each	2,500.00
5,000	Each	5,700.00
10,000	Each	8,200.00
Fuel Storage - Fiberglass - 2,000 Gallons	Each	3,000.00
5,000	Each	5,300.00
10,000	Each	8,700.00
Hot Water Storage - 80 Gallons	Each	360.00
140	Each	550.00
200	Each	760.00
400	Each	1,200.00
600	Each	1,800.00
Liquid Expansion - 30 Gallons	Each	320.00
60	Each	520.00
100	Each	710.00

1502.0 MECHANICAL INSULATION

.1   INSULATION FOR PIPING

	Unit	3/8" Foam	1" Wall Fiberglass	2" Wall Fiberglass
Interior - Pipe 1/2"	LnFt	3.30	4.02	5.70
3/4"	LnFt	3.40	4.27	6.10
1"	LnFt	3.71	4.43	6.20
1-1/4"	LnFt	3.91	4.79	6.40
1-1/2"	LnFt	4.12	5.10	6.80
2"	LnFt	4.53	5.40	7.50
2-1/2"	LnFt	4.89	5.80	8.20
3"	LnFt	5.30	6.30	9.50
6"	LnFt	7.50	7.90	12.90
8"	LnFt	9.20	11.50	14.40
Exterior - w/Aluminum Jacket - Add 2.00/ LnFt				

.2   INSULATION FOR DUCT WORK

	Unit	1"	1-1/2"	2"
Rigid w/ Vapor Barrier	SqFt	2.11	2.94	2.37
Blanket	SqFt	4.43	3.24	2.63

.3   INSULATION FOR BOILERS

	Unit	1"	1-1/2"	2"
Fiberglass - 2"	SqFt	-	-	10.20
Calcium Silicate - 2" with C.F.	SqFt	-	-	15.50

DIVISION #15 - MECHANICAL

		UNIT	COST
1503.0	**FIRE PROTECTION EQUIPMENT**		
.1	AUTOMATIC SPRINKLER EQUIPMENT - Wet System	SqFt	1.87
		(or) Head	182.00
	Dry System	SqFt	2.18
		(or) Head	196.00
	Add for Pendant Type Sprinklers	SqFt	.52
	Add for Deluge System	SqFt	1.82
.2	CARBON DIOXIDE EQUIPMENT - Cylinder	Each	870.00
	Detector & Nozzle	Each	290.00
.3	STANDPIPE AND FIRE HOSE EQUIPMENT		
	Standpipe - Exposed 6"	Each	550.00
	Concealed 6"	Each	700.00
	Cabinets - 30" x 30" Aluminum	Each	290.00
	Hose and Nozzle - 1 1/2"	LnFt	2.60
	2 1/2"	LnFt	3.12
.4	HALON SYSTEM (COMPUTER ROOM)	Each	3,600.00
1504.0	**PLUMBING**		
.1	EQUIPMENT	-	-
.2	PACKAGE WASTE, VENT OR WATER PIPING	Each	410.00
.3	DOMESTIC WATER SOFTENER	Each	770.00
.4	PLUMBING FIXTURES (Drain, Waste and Vent Not Included - See 1504.5)		
	Porcelain Enameled (unless otherwise noted) with accessories		
	Bathtub - 5' - Cast Iron - Enameled	Each	1,000.00
	Steel - Enameled	Each	510.00
	Fiberglass	Each	540.00
	5' x 5' - Cast Iron - Enameled	Each	1,300.00
	Bathtub/ Whirlpool - Fiberglass 6'	Each	3,700.00
	Bidet	Each	890.00
	Drinking Fountain - Enameled Wall Hung	Each	560.00
	SS Recessed	Each	680.00
	Garbage Disposal - 1/2 Horsepower	Each	156.00
	Lavatories - 19" x 17"- Oval	Each	270.00
	20" x 18" - Rectangular	Each	320.00
	27" x 23" - Oval Pedestal	Each	690.00
	18" - SS Bowl	Each	350.00
	Add for Handicap Lavatory	Each	172.00
	Shower & Tub - Combined - Fiberglass	Each	1,200.00
	Shower/Stall - Fiberglass - 32" x 32"	Each	760.00
	Fiberglass - 36" x 36"	Each	1,000.00
	Sink, Kitchen - 24" x 21" - Single	Each	490.00
	24" x 21" - SS Single	Each	350.00
	33" x 22" - Double	Each	540.00
	32" x 21" - SS Double	Each	410.00
	Sink, Laundry - Enameled	Each	540.00
	Plastic	Each	350.00
	Sink, Service - 24" x 21"	Each	660.00
	Urinals - Wall Hung	Each	590.00
	Stall or Trough 4'	Each	790.00
	Wash Fountain - Circular - Precast Terrazzo	Each	1,900.00
	Semi-circular - Precast Terrazzo	Each	1,900.00
	Water Closets - Floor Mounted - Tank Type	Each	430.00
	One Piece	Each	580.00
	Wall Mounted	Each	600.00
	Add for Colored Fixtures	Each	25%
	Add for Handicap Water Closet	Each	94.00
	See 1106 for Appliances		
.5	RESIDENTIAL PLUMBING		
	Average Domestic Fixture, Rough In and Trim	Each	470.00
	Average Cost of Drain, Waste, and Vent	Each	830.00
	Floor Drains - 4"	Each	430.00
.6	POOL EQUIPMENT - See 1318.0	-	-
.7	FOUNTAIN PIPING	-	-

DIVISION #15 - MECHANICAL

		UNIT	COST
1505.0	**HEAT GENERATION**		
.1	FUEL HANDLING EQUIPMENT		-
.2	ASH REMOVAL SYSTEM - Electric	Each	8,100.00
	Hand	Each	5,000.00
.3	LINED BREECHINGS		-
.4	BOILERS		
	Package Type - 1,000 Lbs per Hour	Each	25,000.00
	Steam & Hot Water - 2,000 Lbs per Hour	Each	31,200.00
	5,000 Lbs per Hour	Each	52,000.00
	10,000 Lbs per Hour	Each	57,200.00
	15,000 Lbs per Hour	Each	81,100.00
	20,000 Lbs per Hour	Each	87,400.00
	25,000 Lbs per Hour	Each	117,500.00
	Cast Iron - 100 MBH	Each	2,200.00
	Hot Water or Steam - 200 MBH	Each	3,800.00
	300 MBH	Each	4,800.00
	600 MBH	Each	8,100.00
	1,000 MBH	Each	13,500.00
	Hot Water Heaters - 20 GPM	Each	560.00
	30 GPM	Each	750.00
	40 GPM	Each	810.00
	50 GPM	Each	1,100.00
1506.0	**REFRIGERATION**		
.1	REFRIGERANT COMPRESSORS		-
.2	CONDENSERS - Air Cooled - 10 Ton	Each	13,000.00
	20 Ton	Each	15,800.00
	30 Ton	Each	22,700.00
	40 Ton	Each	27,300.00
	60 Ton	Each	38,900.00
	80 Ton	Each	46,500.00
	Add for Water Cooled	Each	10%
.3	CHILLERS - Hermetic Centrifugal - 100 Ton	Each	50,900.00
	150 Ton	Each	53,600.00
	300 Ton	Each	74,600.00
	Add for Absorption Type	Each	15%
.4	COOLING TOWERS - 100 Ton	Each	7,000.00
	150 Ton	Each	13,000.00
	300 Ton	Each	35,200.00
1507.0	**HEAT TRANSFERS**		
.1	HOT WATER SPECIALTIES		-
.2	CONDENSATE PUMPS AND RECEIVER SETS		
	Outdoor - Air Cooled - 30,000 BTU	Each	1,800.00
	50,000 BTU	Each	2,300.00
	Indoor - 30,000 BTU	Each	900.00
	50,000 BTU	Each	1,200.00
.3	HEAT EXCHANGERS - 100 GPM	Each	1,000.00
	200 GPM	Each	1,100.00
	300 GPM	Each	2,500.00
	600 GPM	Each	5,000.00
	1,000 GPM	Each	6,700.00
.4	TERMINAL UNITS		
	Base Boards - Cast Iron Radiant - 7 1/4"	LnFt	20.80
	9 3/4"	LnFt	22.30
	Tube with Cover - 7 1/4" - 3/4"	LnFt	11.40
	7 1/4" - 1"	LnFt	12.50
	Tube with Cover - 9 3/4" - 3/4"	LnFt	14.10
	9 3/4" - 1"	LnFt	17.40
	14" Fin Tube - Steel	LnFt	21.30
	Copper	LnFt	26.00
	Convectors	Each	250.00
	Radiators	SqFt	7.50

DIVISION #15 - MECHANICAL

		UNIT	COST
1507.0	**HEAT TRANSFERS, Cont'd...**	-	-
.5	COILS		
.6	UNIT HEATERS - 100,000 BTU	Each	640.00
	150,000 BTU	Each	870.00
	200,000 BTU	Each	1,000.00
.7	PACKAGED HEATING AND COOLING	-	2,000.00
.8	STEAM SPECIALTIES	-	450.00
1508.0	**AIR HANDLING AND DISTRIBUTION**		
.1	FURNACE - Duct - 100 MBH	Each	1,150.00
	200 MBH	Each	1,650.00
	300 MBH	Each	2,900.00
.2	FANS - 1,000 CFM	Each	120.00
	5,000 CFM	Each	520.00
	10,000 CFM	Each	860.00
.3	DUCT WORK - Galvanized - 26 Ga	Lb	2.81
	24 Ga	Lb	3.02
	22 Ga	Lb	3.43
	20 Ga	Lb	3.74
.4	DUCT ACCESSORIES		
	Constant Volume Duct	Each	510.00
	Reheat Constant Volume Duct	Each	620.00
	Diffusers & Registers - Sidewalk	SqFt	26.00
	Ceiling	SqFt	44.70
.5	RESIDENTIAL HEATING AND AIR CONDITIONING		
	Furnace	Each	1,300.00
	Central Air Conditioning	Each	620.00
	Humidifier	Each	300.00
	Air Cleaner	Each	490.00
	Thermostat	Each	125.00
	Attic Fan	Each	290.00
	Ventilating Damper	Each	109.00
	6" Flue	Each	330.00
.6	COMMERCIAL AIR CONDITIONING		
	Average	Ton	2,150.00
	High	Ton	3,200.00
	Low	Ton	1,050.00
1509.0	**SEWAGE DRAINAGE**		
.1	SEWAGE EJECTORS 100 GPM	Each	6,900.00
.2	GREASE INTERCEPTORS - 200 GPM	Each	8,100.00
	300 GPM	Each	14,800.00
.3	LIFT STATIONS - STEEL, CONCRETE OR FIBERGLASS		
	WITH GENERATOR - 100 GPM	Each	69,700.00
	200 GPM	Each	117,500.00
	500 GPM	Each	160,200.00
.4	SEPTIC TANKS AND DRAIN FIELD		
	Concrete Tanks with Excav. - 1,000 Gallons	Each	1,400.00
	2,000	Each	2,900.00
	5,000	Each	6,400.00
	10,000	Each	12,000.00
	Drain Fields	LnFt	4.26
.5	SEWAGE TREATMENT EQUIPMENT - STEEL OR CONCRETE		
	5,000 Gallons	GPD	8.80
	50,000	GPD	3.22
	100,000	GPD	3.02
.6	AERATION EQUIPMENT - STEEL - 1,000 Gallons	Each	16,100.00
	CAPACITY PER DAY 2,000	Each	23,500.00
	5,000	Each	32,200.00
	10,000	Each	39,000.00
.7	SLUDGE DIGESTION		-
1510.0	**CONTROLS AND INSTRUMENTATION** - No cost for items listed on Index		-
1511.0	**TESTING, ADJUSTING AND BALANCING**		

DIVISION #16 - ELECTRICAL

Wage Rates (Including Fringes) & Location Modifiers
July 2002-2003

	Metropolitan Area		Electrician Wage Rate	Wage Rate Location Modifier
1.	Akron	*	36.30	104
2.	Albany-Schenectady-Troy	*	34.30	104
3.	Atlanta		29.90	87
4.	Austin	**	28.80	83
5.	Baltimore	*	34.80	99
6.	Birmingham		24.30	72
7.	Boston		47.20	140
8.	Buffalo-Niagara Falls		39.20	111
9.	Charlotte	**	22.80	63
10.	Chicago-Gary	*	46.60	130
11.	Cincinnati	*	29.00	85
12.	Cleveland		49.10	112
13.	Columbus	**	35.60	98
14.	Dallas-Fort Worth	*	29.70	84
15.	Dayton	*	31.80	97
16.	Denver-Boulder	*	30.60	89
17.	Detroit		43.20	120
18.	Flint	*	34.80	100
19.	Grand Rapids	*	34.30	99
20.	Greensboro-West Salem	**	22.40	62
21.	Hartford-New Britain	*	42.20	118
22.	Houston	**	32.90	92
23.	Indianapolis		36.40	106
24.	Jacksonville	**	29.10	83
25.	Kansas City	*	36.30	103
26.	Los Angeles-Long Beach		41.80	128
27.	Louisville	**	29.90	87
28.	Memphis	**	30.60	84
29.	Miami	**	31.50	92
30.	Milwaukee	*	39.00	109
31.	Minneapolis-St. Paul		39.50	115
32.	Nashville	**	29.30	68
33.	New Orleans		25.10	74
34.	New York		61.50	167
35.	Norfolk-Portsmouth	**	26.30	76
36.	Oklahoma City	**	26.10	80
37.	Omaha-Council Bluffs	*	31.20	91
38.	Orlando	**	29.30	88
39.	Philadelphia		47.80	135
40.	Phoenix	*	30.80	95
41.	Pittsburgh		37.90	106
42.	Portland	*	42.10	110
43.	Providence-Pawtucket	*	37.50	109
44.	Richmond	**	26.80	76
45.	Rochester	*	35.20	105
46.	Sacramento	*	46.60	140
47.	St. Louis		43.00	124
48.	Salt Lake City-Ogden	**	31.00	89
49.	San Antonio	**	31.10	82
50.	San Diego	*	44.40	125
51.	San Francisco-Oakland-San Jose	*	51.20	140
52.	Seattle-Everett		39.10	113
53.	Springfield-Holyoke-Chicopee	*	35.60	103
54.	Syracuse	*	37.80	99
55.	Tampa-St. Petersburg	**	30.50	88
56.	Toledo	*	37.10	108
57.	Tulsa	**	28.00	80
58.	Tucson	**	32.50	94
59.	Washington D.C.	*	35.50	88
60.	Youngstown-Warren	*	33.30	97
	AVERAGE		**35.28**	

Impact Ratio: Labor 35%, Material 65%
* Contract Not Settled - Wage Interpolated or Unknown
** Non Signatory or Open Shop Rate

DIVISION #16 - ELECTRICAL

Costs of Electrical Work are priced as total contractors' costs with an overhead and fee of 20% included.

		PAGE
1601.0	**BASIC MATERIALS & METHODS (CSI 16050)**	**16-3**
.1	CONDUITS	16-3
.2	BOXES, FITTINGS AND SUPPORTS	16-3
.3	WIRE AND CABLE	16-3
.4	SWITCHGEAR	16-3
.5	SWITCHES AND RECEPTACLES	16-4
.6	TRENCH DUCT AND UNDERFLOOR DUCT	16-4
.7	MOTOR WIRING	16-4
.8	MOTOR STARTERS	16-4
.9	CABLE TRAY	16-5
1602.0	**POWER GENERATION (CSI 16200)**	**16-5**
1603.0	**HIGH VOLTAGE DISTRIBUTION (CSI 16300)**	**16-5**
1604.0	**SERVICE AND DISTRIBUTION (CSI 16400)**	**16-5**
.1	METERING	16-5
.2	GROUNDING	16-5
.3	SERVICE DISCONNECTS	16-5
.4	DISTRIBUTION SWITCHBOARDS	16-5
.5	LIGHT AND POWER PANELS	16-5
.6	TRANSFORMERS	16-5
1605.0	**LIGHTING (CSI 16500)**	**16-6**
.1	LIGHT FIXTURES	16-6
.2	LAMPS	16-6
1606.0	**SPECIAL SYSTEMS (CSI 16600)**	**16-7**
1607.0	**COMMUNICATION SYSTEMS (CSI 16700)**	**16-7**
.1	TELEPHONE EQUIPMENT	16-7
.2	FIRE ALARM EQUIPMENT	16-7
.3	SECURITY SYSTEMS	16-7
.4	CLOCK PROGRAM EQUIPMENT	16-7
.5	INTER-COMMUNICATION EQUIPMENT	16-7
.6	SOUND EQUIPMENT	16-7
.61	Public Address System (Control and Speakers)	16-7
.62	Nurses and School Call Systems (Panel and Stations)	16-7
.7	TELEVISION ANTENNA SYSTEMS	16-8
.8	CLOSED CIRCUIT TELEVISION STATIONS	16-8
.9	DOCTOR'S REGISTER	16-8
1608.0	**HEATING AND COOLING (CSI 16850)**	**16-8**
.1	SNOW MELTING EQUIPMENT	16-8
.2	ELECTRICAL HEATING EQUIPMENT	16-8
1609.0	**CONTROLS AND INSTRUMENTATION (CSI 16900)**	**16-8**
.1	LIGHTING CONTROL EQUIPMENT	16-8
.2	DIMMING EQUIPMENT	16-8
.3	TEMPERATURE CONTROL EQUIPMENT	16-8
1610.0	**TESTING**	**16-8**
1611.0	**RESIDENTIAL WORK**	**16-8**

DIVISION #16 - ELECTRICAL

1601.0 BASIC MATERIALS AND METHODS

.1 CONDUITS (Incl Fittings/Supports)   COST

.11 Rigid

	UNIT	1/2"	3/4"	1-1/2"	2"	3"	4"	5"	6"
Aluminum	LnFt	4.16	4.47	5.50	7.20	8.80	14.60	20.10	47.80
Galvanized	LnFt	4.37	4.68	5.90	8.40	9.80	19.80	26.50	54.00
EMT (Thinwall)	LnFt	2.86	3.33	3.64	4.89	5.70	10.40	-	-
PVC	LnFt	2.81	2.91	3.07	3.85	4.89	7.90	-	-
Add for Running Exposed									

.12 Flexible

	UNIT	1/2"	3/4"	1-1/2"	2"	3"	4"	5"	6"
Greenfield	LnFt	2.60	2.84	3.43	4.99	7.30	-	-	-
Sealtite	LnFt	3.33	3.48	4.99	7.00	9.40	21.80	-	-

.13 Ducts

	UNIT	1/2"	3/4"	1-1/2"	2"	3"	4"	5"	6"
Fibre	LnFt	-	-	-	-	3.02	3.07	3.85	4.47
Transite	LnFt	-	-	-	-	3.02	3.54	4.26	4.89

Cost for Conduits includes Labor, Material, Equipment and Fees and Installed in Wood or Steel Construction.

.2 BOXES, FITTINGS AND SUPPORTS (In Above)   COST

.3 WIRE AND CABLE

	UNIT	Copper T.H.W.	Aluminum Stranded
Wire - 3 - #14 - 600 Volt	LnFt	3.33	3.00
3 - #12 - 600 Volt	LnFt	3.69	3.30
3 - #10 - 600 Volt	LnFt	4.47	3.90
3 - # 8 - 600 Volt	LnFt	5.70	4.60
3 - # 4 - 600 Volt	LnFt	9.10	6.40
3 - # 2 - 600 Volt	LnFt	15.10	7.80
3 - 1/0	LnFt	20.30	11.00
3 - 2/0	LnFt	23.10	13.00
3 - 3/0	LnFt	32.20	14.50
3 - 4/0	LnFt	38.00	16.00
3 - 250 - MCM	LnFt	43.70	19.00
3 - 300 - MCM	LnFt	49.90	19.50
3 - 350 - MCM	LnFt	55.00	23.50
3 - 400 - MCM	LnFt	62.50	25.50
3 - 500 - MCM	LnFt	67.50	28.50

		ARMORED COPPER		NON-METALLIC	
	UNIT	2 Wire	3 Wire	2 Wire	3 Wire
Cable - #14 bx - 600 Volt	LnFt	1.77	1.98	1.46	1.82
#12 bx - 600 Volt	LnFt	1.87	2.18	1.61	2.29
#10 bx - 600 Volt	LnFt	2.18	2.60	2.08	2.65
# 8 bx - 600 Volt	LnFt	-	3.38	-	3.02
# 6 bx - 600 Volt	LnFt	-	3.74	-	4.00
# 4 bx - 600 Volt	LnFt	-	4.16	-	5.10
Add for Work over 10'					10%
Add for Work in Masonry Construction					10%
Add for Work in Concrete Construction					20%

.4 SWITCH GEAR

Metal Clad 5 KV	22,700.00
15 KV	26,800.00
Add for Outdoor Type	4,350.00

16-3

DIVISION #16 - ELECTRICAL

1601.0 BASIC MATERIALS AND METHODS, Cont'd...

	UNIT	COST
.5 SWITCHES AND RECEPTACLES		
Switches		
Standard Toggle - S P	Each	51.50
4 W	Each	58.00
4 WL	Each	72.00
Mercury - S P	Each	84.50
4 W	Each	155.00
Safety - 30 Amp	Each	165.00
60 Amp	Each	210.00
100 Amp	Each	360.00
200 Amp	Each	470.00
Three-way -	Each	84.50
Add for Dimmer - 600 Watt	Each	79.50
Add for Dimmer - 1500 Watt	Each	170.00
Receptacles		
Single	Each	49.40
Duplex - GFI	Each	68.00
Add for Dedicated - Separate Circuit	Each	129.00
Add for Dedicated - Ground	Each	17.50
Add for Dedicated - Fourplex	Each	67.00
Clock, Lamps, etc.	Each	70.00
Add for Weatherproof	Each	17.50
Plates - Stainless Steel	Each	8.80
Brass	Each	14.00
.6 TRENCH DUCT AND UNDERFLOOR DUCT (Including Fittings)		
Underfloor Duct - Average		
Single Cell - 1 1/4" x 3 1/8"	LnFt	17.70
1 1/2" x 6 1/2"	LnFt	22.40
Double Cell - 1 1/4" x 3 1/8"	LnFt	26.20
1 1/2" x 6 1/2"	LnFt	43.70
Wireway - Cover Type 2 1/2" x 2 1/2" x 1"	Each	18.80
2 1/2" x 2 1/2" x 2"	Each	25.00
2 1/2" x 2 1/2" x 3"	Each	31.20
4" x 4" x 1"	Each	22.90
4" x 4" x 2"	Each	29.10
4" x 4" x 3"	Each	37.40
.7 MOTOR WIRING (Including Controls)		
10 H.P.	Each	3,750.00
20 H.P.	Each	3,850.00
30 H.P.	Each	4,350.00
40 H.P.	Each	4,600.00
50 H.P.	Each	5,500.00
100 H.P.	Each	7,300.00
200 H.P.	Each	11,100.00
.8 MOTOR STARTERS (600 Volt)		
Magnetic - 10 H.P.	Each	400.00
Non-Reversing - 20 H.P.	Each	560.00
30 H.P.	Each	820.00
40 H.P.	Each	950.00
50 H.P.	Each	1,100.00
100 H.P.	Each	2,200.00
200 H.P.	Each	4,250.00
Add for Reversing Type	Each	80%
Add for Explosion Proof	Each	40%
Add for Motor Circuit Protectors	Each	25%
Add for Fused Switches	Each	10%
Manual - 1 H.P.	Each	101.00
3 H.P.	Each	166.00
7 1/2 H.P.	Each	198.00

DIVISION #16 - ELECTRICAL

1601.0 BASIC MATERIALS AND METHODS, Cont'd...

		UNIT	COST
.9	CABLE TRAY (Aluminum or Galvanized)		
	Ladder or Punched - 6" wide	LnFt	15.60
	12" wide	LnFt	16.60
	18" wide	LnFt	18.70
	24" wide	LnFt	20.80

1602.0 POWER GENERATION

		UNIT	COST
	Emergency Generators (3-Phase, 4-Wire) (120/208 Volt)		
	15 k.w.	Each	10,900.00
	30 k.w.	Each	16,600.00
	40 k.w.	Each	20,800.00
	50 k.w.	Each	27,600.00
	75 k.w.	Each	3,200.00
	100 k.w.	Each	38,500.00
	300 k.w.	Each	72,800.00

1603.0 HIGH VOLTAGE DISTRIBUTION - Power Companies

1604.0 SERVICE AND DISTRIBUTION

		UNIT	COST
.1	METERING		
.2	GROUNDING (Including Rods and Accessories)		
	1/2" x 10'	Each	62.50
	5/8" x 10'	Each	75.00
	3/4" x 10'	Each	88.50
.3	SERVICE DISCONNECTS - 100 amp	Each	310.00
	200 amp	Each	470.00
	400 amp	Each	915.20
	600 amp	Each	1,800.00
.4	DISTRIBUTION SWITCHBOARDS (240 Volt) - 400 amp	Each	3,200.00
	800 amp	Each	3,650.00
	1600 amp	Each	5,300.00
.5	LIGHT AND POWER PANELS (3-Phase, 4-Wire) (120/208 Volt)		
	12 circuit	Each	750.00
	16 circuit	Each	880.00
	20 circuit	Each	1,050.00
	24 circuit	Each	1,250.00
	30 circuit	Each	1,450.00
	36 circuit	Each	1,600.00
	Add for Circuit Breakers	Each	230.00
.6	TRANSFORMERS		
	Current - 400 amp	Each	3,100.00
	600 amp	Each	3,500.00
	800 amp	Each	3,950.00
	1000 amp	Each	4,350.00

Light and Power	UNIT	SINGLE-PHASE 120/ 240 Volt	3-PHASE 120/ 208 Volt
Dry Type - 3 kva	Each	360.00	530.00
5 kva	Each	540.00	780.00
10 kva	Each	750.00	1,000.00
15 kva	Each	1,050.00	1,400.00
30 kva	Each	1,650.00	1,850.00
45 kva	Each	2,400.00	2,450.00
75 kva	Each	3,350.00	3,200.00
Oil Filled - 150 kva	Each	-	5,600.00
300 kva	Each	-	10,600.00
500 kva	Each	-	13,800.00

DIVISION #16 - ELECTRICAL

1605.0 LIGHTING

	UNIT	COST
.1 LIGHT FIXTURES		
Fluorescent		
Surface		
Industrial - 48" - 1 lamp R.S.	Each	85.50
H.O.	Each	124.00
2 lamp R.S.	Each	127.00
H.O.	Each	158.00
4 lamp R.S.	Each	139.00
H.O.	Each	240.00
96" - 1 lamp H.O.	Each	134.00
R.S.	Each	240.00
2 lamp H.O.	Each	118.00
H.O.	Each	240.00
Commercial - 12" x 48" - 2 lamp R.S.	Each	134.00
24" x 24" - 4 lamp R.S.	Each	185.00
24" x 48" - 4 lamp R.S.	Each	250.00
48" x 48" - 4 lamp R.S.	Each	370.00
12" x 48" - 2 lamp R.S.	Each	172.00
Recessed - 24" x 24" - 4 lamp R.S.	Each	210.00
24" x 48" - 4 lamp R.S.	Each	240.00
48" x 48" - 4 lamp R.S.	Each	410.00
Add for Electronic Ballasts	Each	22.70
Incandescent - 100 watt	Each	92.50
200 watt	Each	108.00
300 Watt	Each	124.00
Mercury Vapor - 250 watt Single	Each	440.00
400 watt Single	Each	480.00
Twin	Each	620.00
Quartz - 500 watt	Each	165.00
Sodium - Low Pressure - 35 watt	Each	300.00
55 watt	Each	440.00
Explosion Proof		
Fluorescent - 48" - 2 lamp R.S.	Each	1,050.00
Incandescent - 48" - 200 watt	Each	270.00
Mercury Vapor - 48" - 200 watt	Each	660.00
Emergency - Nickel Cadmium Battery	Each	710.00
Lead Battery	Each	360.00
Exit - Standard	Each	108.00
Exit - Low Voltage - Battery	Each	185.00
Porcelain	Each	62.00
.2 LAMPS		
Fluorescent - 48" - 40 watt	Each	6.00
48" - 60 watt	Each	7.70
96" - 75 watt	Each	10.80
Incandescent - 25 to 100 watt	Each	2.16
400 watt	Each	3.35
Mercury Vapor - 100 watt	Each	43.30
250 watt	Each	54.50
400 watt	Each	74.00
1000 watt	Each	95.00
Metal Halide - 175 watt	Each	65.00
400 watt	Each	70.00
1000 watt	Each	144.00
Sodium - 300 watt	Each	87.50
400 watt	Each	95.00
1000 watt	Each	220.00
Add for Disposal	Each	1.03

DIVISION #16 - ELECTRICAL

			UNIT	COST
1606.0	**SPECIAL SYSTEMS**			
	LIGHTNING PROTECTION			
		Ground Rod and Copper Point - 3/4" to 10"	Each	85.50
		Ground Clamps - 2"	Each	104.00
		Wire - #2	LnFt	1.14
1607.0	**COMMUNICATION SYSTEMS**			
.1	TELEPHONE EQUIPMENT			
		Cabinets 24" x 36"	Each	310.00
		36" x 48"	Each	470.00
		Lines	Each	280.00
		Phones	Each	160.00
		Station	Each	590.00
.2	FIRE ALARM EQUIPMENT			
		Control Panel 10 zones	Each	1,450.00
		20 zones	Each	2,450.00
		Annunciator Panel 10 zones	Each	490.00
		Station	Each	160.00
		Horn	Each	103.00
		Heat Detector	Each	72.00
		Smoke Detector	Each	149.50
.3	SECURITY SYSTEMS			
		Standard Contact	Each	210.00
		Audible Alarm	Each	280.00
		Photo Electric Sensor - 500' Range	Each	440.00
		Monitor Panel - Standard	Each	490.00
		High Security	Each	680.00
		Vibration Sensor	Each	185.00
		Audio Sensor	Each	134.00
		Motion Sensor	Each	320.00
.4	CLOCK PROGRAM EQUIPMENT (Control Panel and Clock with Buzzer)			
		Master Time Clock 10 rm - average	Each	340.00
		20 rm - average	Each	300.00
		30 rm - average	Each	260.00
		Time Clock and Cards	Each	1,100.00
		Add for Electronic and Bells	Each	20%
.5	INTER-COMMUNICATION EQUIPMENT			
		Individual Stations	Each	134.00
		Board - Master	Each	520.00
.6	SOUND EQUIPMENT			
.61	Public Address System (Control and Speakers)			
		3 to 6 Speakers - average each speaker	Each	740.00
		7 to 9 Speakers - average each speaker	Each	680.00
		10 to 20 Speakers - average each speaker	Each	630.00
		Add for Tape / CD	Each	490.00
.62	Nurses & School Call systems (Panel & Stations)			
		10 Stations or Rooms - average each room	Each	680.00
		20 Stations or Rooms - average each room	Each	630.00
		30 Stations or Rooms - average each room	Each	540.00
		Add per Light	Each	67.00

DIVISION #16 - ELECTRICAL

		UNIT	COST
1607.0	**COMMUNICATIONS SYSTEMS, Cont'd...**		
.7	TELEVISION ANTENNA SYSTEMS		
	School Systems - 10 outlets	Each	320.00
	20 outlets	Each	310.00
	30 outlets	Each	290.00
	Apartment Houses - 20 outlets	Each	139.00
	30 outlets	Each	129.00
	50 outlets	Each	124.00
.8	CLOSED CIRCUIT T.V. STATIONS		
	Recorder	Each	3,650.00
	Camera	Each	1,450.00
	Surveillance	Each	1,900.00
.9	DOCTOR'S REGISTER	Each	6,700.00
	Add for Recording Register	Each	5,800.00
1608.0	**HEATING AND COOLING**		
.1	SNOW MELTING EQUIPMENT - 480 volt	SqFt	9.60
.2	ELECTRIC HEATING EQUIPMENT		
	Baseboard - 500 watt	LnFt	35.40
	750 watt	LnFt	34.30
	1000 watt	LnFt	33.30
	Cable Heating	SqFt	15.60
	Unit Heaters - 1500 watt	Each	210.00
	2500 watt	Each	310.00
1609.0	**CONTROLS AND INSTRUMENTATION**		
.1	LIGHTING CONTROL EQUIPMENT		-
.2	DIMMING EQUIPMENT		-
.3	TEMPERATURE CONTROL EQUIPMENT		-
1610.0	**TESTING**		-
1611.0	**RESIDENTIAL WORK**		
	100-Amp Service	Each	800.00
	200-Amp Service - Single Phase	Each	1,250.00
	Three Phase	Each	2,300.00
	Air Conditioner	Each	260.00
	Dimmer - 600 Watt	Each	52.00
	Dishwasher	Each	73.00
	Doorbell	Each	52.00
	Dryer - Electric	Each	104.00
	Dryer or Washer - Gas	Each	73.00
	Duplex Receptacle	Each	26.00
	Dedicated	Each	60.50
	Exhaust Fan - Bath	Each	49.90
	Kitchen	Each	59.50
	Fixture, Average	Each	88.50
	Porcelain	Each	83.00
	Furnace	Each	88.50
	Garbage Disposal	Each	73.00
	Paddle Fan	Each	156.00
	Phone - Data Box	Each	41.60
	Range Circuit	Each	125.00
	Smoke Detectors	Each	59.50
	Switches - Single	Each	26.00
	3-Way	Each	60.50
	Thermostat	Each	43.70
	Water Heater - 20 Amp 120 Volt	Each	114.00
	Whirlpool	Each	109.00

APPENDIX

Average Square Foot Costs
2003

Housing (with Wood Frame)

Square Foot Costs do not include Garages, Land, Furnishings, Equipment, Landscaping, Financing, or Architect's Fee. Total Costs, except General Contractor's Fee, are included in each division. See Page 3 for Metropolitan Cost Variation Modifier. See Division 13 for Metal Buildings, Greenhouses, and Air-Supported Structures.

	Division	Houses - 3 Bedrooms with Basement			Houses - 3 Bedrooms without Basement		
		Low Rent	Project	Custom	Low Rent	Project	Custom
1.	General Conditions	$3.70	$3.75	$4.05	$4.85	$4.90	$5.30
2.	Site Work	2.45	2.70	3.00	4.15	4.30	4.35
3.	Concrete	3.20	3.35	3.50	5.80	6.00	6.50
4.	Masonry	4.20	4.45	5.85	5.80	6.20	8.10
5.	Steel	1.05	1.05	1.15	.50	.50	.50
6.	Carpentry	12.90	13.40	14.20	18.50	18.70	20.80
7.	Moisture & Therm. Prot.	3.20	3.20	3.40	4.50	4.65	5.00
8.	Doors, Windows & Glass	6.40	6.70	7.00	8.50	8.70	9.10
9.	Finishes	6.00	6.50	7.90	8.80	8.90	12.60
10.	Specialties	.65	.65	.70	.70	.70	.95
11.	Equipment (Cabinets)	2.35	2.40	3.50	4.15	5.10	6.20
12.	Furnishings	.45	.45	.50	.75	.75	.75
13.	Special Construction	-	-	-	-	-	-
14.	Conveying	-	-	-	-	-	-
15.	Mechanical: Plumbing	4.50	4.55	5.30	6.70	6.70	8.30
	Heating	3.10	3.20	3.30	4.35	4.60	6.50
	Air Cond.	-	-	-	-	-	-
16.	Electrical	2.65	2.70	3.25	4.00	4.00	4.80
	SUB TOTAL	$56.80	$59.05	$66.30	$82.05	$84.70	$99.75
	CONTRACTOR'S SERVICES	5.70	5.90	6.80	8.20	8.50	10.00
	TOTAL AVERAGE Sq.Ft. Cost	$62.50	$64.95	$73.10	$90.25	$93.20	$109.75

Unfinished Basement, Storage and Utility Areas included in Sq.Ft. Costs:

No Basement Utilities In Finished Area

	With Basement			Without Basement		
Average Cost/ Unit	$125,000	$129,800	$182,500	$108,240	$111,840	$164,100
Average Sq.Ft.Size/ Unit	2,000	2,000	2,500	1,200	1,200	1,500
Add Architect's Fee	$6,250.00	$7,000.00	$16,000.00	$5,400.00	$6,100.00	$12,000.00

	COST	
	Sq. Ft.	Ea. Unit
Add for Finishing Basement Space For Bedrooms or Amusement Rooms, 12' x 20'	$24.00	$5,800
Add for Single Garage (no Interior Finish), 12' x 20'	$23.00	$5,500
Add for Double Garage (no Interior Finish), 20 x 20'	$24.00	$9,700
Add (or Deduct) for Bedrooms, 12' x 12'	$35.00	$5,100
Add per Bathroom with 2 Fixtures, 6' x 8'	$110.00	$5,300
Add per Bathroom with 3 Fixtures, 8' x 8'	$110.00	$7,000
Add for Fireplace	-	$5,700
Add for Central Air Conditioning	-	$4,500
Add for Drain Tiling	-	$2,000
Add for Well	-	$4,600
Add for Sewage System	-	$5,000

APPENDIX

Average Square Foot Costs
2003

Housing

Square Foot Costs do not include Land, Furnishings, Equipment, Landscaping and Financing.

		Apartments			Motels & Hotels	
Division	2 & 3 Story	2 & 3 Story	High Rise	High Rise	1 & 2 Story	High Rise
	(1)a	(1)b	(2)	(3)	(1)b	(2)
1. General Conditions	$ 3.55	$ 3.80	$ 5.30	$ 5.75	$ 3.90	$ 4.95
2. Site Work	2.20	2.15	3.10	3.10	4.50	3.15
3. Concrete	1.80	1.80	17.25	12.50	2.15	17.50
4. Masonry	1.45	8.90	2.35	2.25	8.95	11.50
5. Steel	1.05	1.05	1.65	1.75	1.30	1.60
6. Carpentry and Millwork	13.00	11.30	4.30	4.00	11.20	3.45
7. Moisture Protection	3.00	3.20	1.60	1.60	2.90	1.65
8. Doors, Windows and Glass	4.90	5.35	5.95	6.80	6.20	7.35
9. Finishes	9.10	9.20	9.35	8.80	9.10	12.65
10. Specialties	.70	.70	.75	.75	.90	.90
11. Equipment	1.50	1.95	1.50	1.50	2.45	2.10
12. Furnishings	.75	.80	.85	.85	1.80	1.35
13. Special Construction	-	-	-	-	-	-
14. Conveying	-	2.10	3.10	3.30	-	3.75
15. Mechanical:						
Plumbing	5.50	5.50	5.40	5.30	6.55	8.40
Heating and Ventilation	4.80	5.20	5.75	5.50	5.20	6.50
Air Conditioning	1.70	1.75	1.80	1.80	2.25	4.25
Sprinklers	-	-	1.80	1.65	1.75	1.75
16. Electrical	5.70	5.70	6.60	6.95	8.10	8.20
SUB TOTAL	$60.70	$70.45	$78.40	$74.15	$79.30	$101.00
CONTRACTOR'S SERVICES	6.10	7.05	7.85	7.45	7.90	10.00
Construction Sq.Ft. Cost	$66.80	$77.50	$86.25	$81.60	$87.10	$111.00
Add for Architect's Fee	4.70	6.20	7.25	8.20	7.00	8.80
TOTAL Sq.Ft. Cost	$71.50	$83.70	$93.50	$89.80	$94.10	$119.80
Average Cost per Unit without Architect's Fee	$63,460	$73,625	$77,620	$65,280	$56,620	$77,870
Average Cost per Unit with Architect's Fee	$67,925	$79,515	$83,250	$71,840	$61,100	$77,870

Add for Porches - Wood - $9.50 SqFt
Add for Interior Garages - $8.50 SqFt

Average Sq.Ft. Size per Unit	950	950	900	800	650	650
Average Sq.Ft. Living Space	750	750	750	600	450	450

(1)a Wall Bearing Wood Frame & Wood or Aluminum Facade
(1)b Wall Bearing Masonry & Wood Joists & Brick Facade
(2) Concrete Frame
(3) Post Tensioned Concrete Slabs, Sheer Walls, Window Wall Exteriors, and Drywall Partitions.

APPENDIX

Average Square Foot Costs
2003

Commercial

Square Foot Costs do not include Land, Furnishings, and Financing.

	Division	Remodeling Office Interior	Office 1-Story (1)	Office 2-Story (2)	Office Up To 5-Story (3)	High Rise (City) Above 5-Story (4)	Parking Ramp (3)
1.	General Conditions	$ 2.25	$ 3.65	$ 3.65	5.40	5.70	4.00
2.	Site Work & Demolition	1.35	1.85	1.85	1.85	1.65	1.20
3.	Concrete	-	5.40	8.55	18.50	6.45	20.50
4.	Masonry	-	9.10	8.55	13.00	3.20	1.30
5.	Steel	-	-	7.80	2.80	2.00	1.40
6.	Carpentry and Millwork	5.60	3.00	2.80	4.40	4.85	1.00
7.	Moisture Protection	-	5.40	2.70	1.85	2.30	2.40
8.	Doors, Windows and Glass	1.50	5.60	5.60	11.60	21.60	1.15
9.	Finishes & Sheet Rock	9.60	10.05	10.00	11.80	12.70	1.55
10.	Specialties	-	1.45	1.45	1.50	1.50	-
11.	Equipment	-	4.50	4.00	3.30	4.45	-
12.	Furnishings	-	.95	1.15	1.25	1.35	-
13.	Special Construction	-	-	-	-	-	-
14.	Conveying	-	-	2.55	4.25	4.95	2.55
15.	Mechanical:						
	Plumbing	2.75	4.35	4.55	5.30	5.50	1.80
	Heating and Ventilation	1.50	5.50	5.25	6.85	6.85	-
	Air Conditioning	1.45	5.45	5.30	6.65	6.60	-
	Sprinklers	-	1.70	1.70	1.80	1.80	-
16.	Electrical	5.60	6.85	7.55	8.30	9.55	2.15
	SUB TOTAL	$31.60	$74.80	$85.00	$110.40	$ 103.00	$41.00
	CONTRACTOR'S SERVICES	3.15	7.50	8.50	11.00	10.30	4.10
	Construction Sq.Ft. Cost	$34.75	$82.30	$93.50	$121.40	$113.30	$45.10
	Add for Architect's Fee	3.50	6.40	7.20	8.40	9.00	3.40
	TOTAL Sq.Ft. Costs	**$38.25**	**$88.70**	**$100.70**	**$129.80**	**$122.30**	**$48.50**
	Deduct for Tenant Areas						
	Unfinished	-	-	21.00	23.00	27.00	-
	Per Car Average	-	-	-	-	-	$12,800
	Add: Masonry Enclosed Ramp	-	-	-	-	-	4.80
	Add: Heated Ramp	-	-	-	-	-	4.60
	Add: Sprinklers for Closed Ramps	-	-	-	-	-	1.80
	Deduct for Precast Concrete Ramp	-	-	-	-	-	7.30

(1) Wall Bearing Masonry and Steel Joists and Decks - No Basement
(2) Wall Bearing Masonry and Precast Concrete Decks - No Basement
(3) Concrete Frame
(4) Steel Frame and Window Wall Facade

APPENDIX

Average Square Foot Costs
2003

Commercial and Industrial

Square Foot Costs do not include Land, Furnishings, Landscaping, and Financing.
See Division 13 for Metal Buildings

	Division	Store 1-Story (1)	Store 2-Story (2)	Production 1-Story (1)	Warehouse 1-Story (1)	Warehouse and 1-Story Office (1)	Warehouse and 2-Story Office (3)
1.	General Conditions	$ 3.35	$ 3.35	$ 3.25	$ 2.90	$ 3.00	$ 2.70
2.	Site Work & Demolition	1.70	1.30	1.50	1.45	1.75	1.40
3.	Concrete	4.90	8.90	4.90	4.80	5.30	8.15
4.	Masonry	7.10	7.70	5.95	6.00	6.75	3.65
5.	Steel	7.85	4.15	7.90	7.65	8.30	6.30
6.	Carpentry	2.30	2.35	1.50	1.45	2.00	1.80
7.	Moisture Protection	3.90	2.25	4.25	4.05	4.10	2.05
8.	Doors, Windows and Glass	4.55	4.35	3.60	1.90	3.75	3.35
9.	Finishes	5.20	4.65	4.45	1.60	2.75	4.25
10.	Specialties	1.40	1.40	1.05	1.00	1.00	1.00
11.	Equipment	1.30	1.50	1.40	-	.75	.90
12.	Furnishings	-	-	-	-	-	-
13.	Special Construction	-	-	-	-	-	-
14.	Conveying	-	3.10	-	-	-	-
15.	Mechanical:						
	Plumbing	3.40	3.10	3.60	2.05	2.60	2.40
	Heating and Ventilation	3.60	3.55	2.75	2.85	3.80	3.45
	Air Conditioning	3.85	4.00	2.60	-	1.00	.90
	Sprinklers	1.70	1.65	1.60	1.65	1.70	1.65
16.	Electrical	6.40	6.00	6.70	4.00	4.15	3.75
	SUB TOTAL	$62.50	$63.30	$57.00	$42.40	$52.70	$48.20
	CONTRACTOR'S SERVICES	6.30	6.30	5.70	4.30	5.30	4.80
	Construction Sq.Ft. Cost	$68.80	$69.60	$62.70	$46.70	$57.70	$53.00
	Add for Architect's Fees	5.00	4.50	4.50	3.40	4.10	3.80
	TOTAL Sq.Ft. Cost	$73.85	$74.10	$67.20	$49.70	$61.80	$56.80

(1) Wall Bearing Masonry and Steel Joists and Decks - Brick Fronts
(2) Wall Bearing Masonry and Precast Concrete Decks - Brick Fronts
(3) Precast Concrete Wall Panels, Steel Joists, and Deck

SF-4

APPENDIX

Average Square Foot Costs
2003

Educational and Religious

Square Foot Costs do not include Land, Furnishings, Landscaping and Financing.

	Division	Elementary School 1-Story (1)	High School 2-Story (1)	Vocational School 3-Story (1)	College Multi-Story (2)	Gym (3)	Dorm 3-Story (1)	Church (3)
1.	General Conditions	$5.70	$5.60	$5.65	$5.55	$4.35	$5.10	5.50
2.	Site Work	3.00	3.20	3.00	3.00	1.50	1.95	2.85
3.	Concrete	5.00	5.20	3.75	17.80	5.20	3.95	4.75
4.	Masonry	11.40	10.80	11.20	11.60	11.40	10.70	23.20
5.	Steel	7.90	9.70	9.70	3.30	9.75	8.30	1.45
6.	Carpentry	3.30	3.15	2.85	2.05	1.80	2.80	6.05
7.	Moisture Protection	4.45	3.00	3.40	3.50	4.70	4.05	4.85
8.	Doors-Windows-Glass	5.50	5.40	5.70	7.20	5.10	4.65	6.65
9.	Finishes	9.05	7.50	8.40	7.50	6.45	8.75	9.10
10.	Specialties	1.45	1.40	1.40	1.75	1.40	1.40	1.05
11.	Equipment	4.00	3.85	5.80	5.05	7.60	4.95	6.85
12.	Furnishings	-	.80	1.00	1.05	-	1.00	4.00
13.	Special Construction	-	-	-	-	-	-	-
14.	Conveying	-	-	2.85	2.85	-	2.85	-
15.	Mechanical:							
	Plumbing	6.75	6.90	6.90	9.35	4.45	6.35	4.35
	Heating-Ventilation	10.30	9.90	11.30	10.25	6.55	6.95	7.10
	Air Conditioning	-	-	-	-	-	-	5.50
	Sprinklers	1.75	1.75	1.75	1.75	1.75	1.75	-
16.	Electrical	9.85	10.55	11.10	11.25	7.00	8.10	7.05
	SUB TOTAL	$89.40	$88.70	$95.70	$104.80	$79.00	$83.60	100.30
	CONTRACTOR'S SERVICES	8.90	8.90	9.60	10.50	7.90	8.40	10.00
	Constr. Sq.Ft. Cost	$98.30	$97.60	$105.30	$115.30	$86.90	$92.00	110.30
	Add Architect's Fees	6.90	6.90	7.40	8.10	6.10	6.50	8.00
	TOTAL Sq.Ft. Cost	$105.20	$104.50	$112.70	$123.40	$93.00	$98.50	$118.10

(1) Wall Bearing Masonry and Steel Joists and Decks
(2) Concrete Frame and Masonry Facade
(3) Masonry Walls and Wood Roof

APPENDIX

Average Square Foot Costs
2003

Medical and Institutional

Square Foot Costs do not include Land, Landscaping, Furnishings, and Financing.

		Clinic 1-Story (1)	Doctors Office 1-Story (1)	Hospital 1-Story (1)	Hospital Multi-Story (2)	Nursing Home 1-Story (1)	Housing for Elderly 1-Story (1)
1.	General Conditions	$6.00	$6.10	$8.15	$8.20	$7.35	$6.75
2.	Site Work	1.85	1.85	2.90	2.95	3.10	3.20
3.	Concrete	6.10	6.05	6.20	19.70	6.40	6.50
4.	Masonry	10.60	9.50	17.20	12.00	12.25	10.90
5.	Steel	6.90	6.85	8.40	1.85	1.90	1.60
6.	Carpentry	5.70	5.60	4.90	3.20	3.40	3.25
7.	Moisture Protection	4.30	4.20	4.55	1.95	4.55	4.55
8.	Doors, Windows and Glass	6.90	7.25	9.10	6.10	7.95	6.40
9.	Finishes	9.10	8.90	11.50	11.20	11.20	9.30
10.	Specialties	1.30	1.30	1.40	1.35	1.35	1.30
11.	Equipment	7.00	5.60	10.70	10.55	5.30	2.30
12.	Furnishings	1.05	1.05	1.10	1.20	1.00	2.10
13.	Special Construction	.75	.75	1.55	1.55	.75	.75
14.	Conveying	-	-	-	5.80	-	5.40
15.	Mechanical:						
	Plumbing	9.10	9.75	15.00	14.30	9.80	9.45
	Heating and Ventilation	8.30	7.40	10.90	11.20	5.70	4.50
	Air Conditioning	6.40	6.15	11.20	9.70	6.45	2.45
	Sprinklers	1.75	1.75	1.75	1.75	1.95	1.75
16.	Electrical	11.90	10.20	18.70	16.80	10.90	9.50
	SUB TOTAL	$105.00	$100.20	$145.20	$141.30	$101.30	$92.00
	CONTRACTOR'S SERVICES	10.50	10.00	14.50	14.20	10.20	9.20
	Construction Sq.Ft. Cost	$115.50	$110.20	$159.75	$155.50	$111.50	$101.20
	Add for Architect's Fees	8.00	7.80	11.30	11.00	7.80	7.10
	TOTAL Sq.Ft. Cost	**$123.50**	**$118.00**	**$170.00**	**$165.60**	**$119.30**	**$108.30**
	Average Cost per Bed			$112,000	$111,000	$46,000	
	Average Cost per Patient Room			$162,000	$161,000	$77,000	

Average Cost to Remodel Approximately 50% Cost of New

(1) Wall Bearing Masonry and Steel Joists and Decks
(2) Concrete Frame and Masonry Façade

APPENDIX

Average Square Foot Costs
2003

Government and Finance

Square Foot Costs do not include Land, Landscaping, Furnishings, and Financing.

		Bank & Drive-Thru 1-Story (1)	Bank & Insur. Bldg High-Rise (3)	Govt. Office Bldg Multi-Story (2)	Post Office 1-Story (1)	Court House 3-Story (2)	Jail and Prison (2)
1.	General Conditions	$6.60	$7.60	$7.60	$6.20	$7.50	$8.30
2.	Site Work	3.50	3.20	2.70	2.25	2.45	3.25
3.	Concrete	8.40	20.30	21.40	9.15	18.20	25.20
4.	Masonry	12.60	8.75	14.50	8.70	14.35	16.00
5.	Steel	3.70	1.70	2.10	2.90	1.75	4.20
6.	Carpentry	3.80	3.40	3.50	3.05	2.20	3.90
7.	Moisture Protection	3.15	3.60	3.95	3.15	4.00	4.60
8.	Doors, Windows and Glass	8.90	20.80	9.15	7.05	6.55	8.50
9.	Finishes	12.70	13.80	14.80	12.60	11.95	10.20
10.	Specialties	1.40	1.40	2.70	1.45	2.35	5.60
11.	Equipment	11.00	7.30	7.30	9.70	9.20	10.80
12.	Furnishings	1.25	1.15	-	1.70	-	-
13.	Special Construction	-	-	-	-	-	-
14.	Conveying	-	2.35	3.00	1.55	2.20	2.95
15.	Mechanical						
	Plumbing	5.10	6.45	6.30	6.85	8.70	15.00
	Heating and Ventilation	5.00	6.25	6.80	4.90	7.25	12.90
	Air Conditioning	4.10	4.60	5.75	5.85	10.30	10.75
	Sprinklers	1.80	1.75	1.75	1.75	1.75	1.75
16.	Electrical	11.00	13.20	13.70	10.40	13.00	21.20
	SUB TOTAL	$104.00	$127.60	$127.00	$99.20	$123.70	$165.10
	CONTRACTOR'S SERVICES	10.40	12.80	12.70	10.00	12.40	16.50
	Construction Sq.Ft. Cost	$114.40	$140.40	$139.70	$109.20	$136.10	$181.10
	Add for Architect's Fees	8.00	10.40	11.10	8.70	10.90	14.40
	TOTAL Sq.Ft. Cost	$122.40	$150.80	$150.80	$117.90	$147.00	$195.50

(1) Wall Bearing Masonry and Precast Concrete Decks
(2) Concrete Frame and Masonry Facade
(3) Concrete Frame and Window Wall

APPENDIX

Construction Industry Statistical Data
Number of Employees in Contract Construction Industry
And Total U.S. Labor Force 2001 (2000 Statistics)

1. **Total Population Employment Status** — Total 2000 Employees[1]

Category	Total 2000 Employees[1]
Mining	557,000
Contract Construction (Breakdown below)	6,840,000
Manufacturing	17,939,000
Transportation and Public Utilities	7,096,000
Wholesale Trade	7,066,000
Retail Trade	23,331,000
Finance, Insurance and Real Estate	7,716,000
Service and Miscellaneous	40,644,000
Government - Federal	2,614,000
State	4,807,000
Local	13,217,000
Total Non-Agricultural	**132,170,000**
Agriculture	3,163,000
Total Employed	**135,333,000**
Unemployed	3,326,000
Total Labor Force	**138,659,000**

2. **Contract Construction Employment Status 2000**

	Total Construction[2] Employment	Workers	Architects Engineers Managers Supervision Clerical
General Building Contractors			
Residential - Contractors	791,300	510,300	281,000
Operative - Developers	30,600	12,500	18,100
Non Residential - Contractors	654,800	468,700	186,100
Total General Building	**1,476,700**	**991,500**	**485,200**
Special Trade Contractors			
Plumbing, Heating & Air Conditions	934,900	688,000	246,100
Painting and Decorating	217,900	175,900	42,000
Electrical Work	863,300	677,100	186,200
Masonry, Plastering, Tile and Stone	543,200	469,800	73,400
Carpentry and Flooring	314,900	234,900	80,000
Roofing and Sheet Metal	237,400	179,700	57,700
Other Miscellaneous	1,071,000	781,800	289,300
Total Special Trade	**4,182,600**	**3,207,000**	**974,700**
Heavy Construction			
Highway and Street	228,000	177,000	51,000
Other Heavy	590,700	498,200	92,500
Total Heavy Construction	**818,700**	**675,200**	**143,500**
Total Construction Employees	**6,478,000**	**4,873,900**	**1,603,400**

---

(1) *Employment and Earnings May 2001* - U.S. Department of Labor, Bureau of Labor Statistics
(2) AC&E Projections from 1998 Construction Review which ceased publication in 1998.

APPENDIX

Construction Industry Statistical Data
2001 (2000 Statistics)
Gross Earnings and Hours

	United States[1]			Metropolitan Areas[2]	
	Average Hourly Earnings	Average Weekly Earnings	Gross Hours	Hourly Earnings	40-Hour Weekly Earnings
Mining	17.35	792.90	45.7		
Contract Construction	18.24	693.12	38.0	28.10	1,124.00
Manufacturing	14.79	585.68	39.8		
Transportation and Public Utilities	16.69	624.47	38.6		
Wholesale Trade	15.81	584.43	38.6		
Retail Trade	9.77	282.35	28.9		
Finance, Insurance and Real Estate	15.85	584.87	36.9		
Services	14.57	478.44	32.7		
Agriculture					
Average	13.74	474.03	34.5		
Contract Construction					
General Building Contractors	18.29	702.34[2]	37.8	26.80[3]	1072.00
Residential	16.72	613.62	36.7		
Operative	19.53	720.66	36.9		
Non Residential	18.88	740.10	39.2		
Special Trade Contractors					
Plumbing, Heating and Air Cond.	19.20	748.88	38.0	32.00	1,280.00
Painting and Decorating	16.18	580.88	38.9	27.80	1,112.00
Electrical Work	20.01	820.28	35.9	33.20	1,328.00
Masonry, Plastic, Tile and Stone	18.12	652.32	36.6	29.60	1,184.00
Roofing and Sheet Metal	15.91	528.08	36.6	26.00	1,040.00
Carpentry and Flooring	18.52	675.98	33.1	29.00	1,160.00
Heavy Construction					
Highway and Street	17.12	679.66	40.9		
Other Heavy	17.45	719.88	41.3		
Combined Construction					
Average (39.1 hours)	18.29	702.34	38.4	28.10	1,098.00
Average (40.0 hours)				28.10	1,124.00
[2] Skilled 80%				29.50	1,180.00
Laborers 20%				22.50	900.00
Weighted Skilled and Laborers				28.10	1,124.00

---

[1] *Employment and Earnings May 2001 -*
U.S. Department of Labor, Bureau of Labor Statistics

[2] A.C.& E. 60 Metropolitan Areas - Weighted Index
   Skilled Average = 28.00
      Plumbers & Sheet Metal - 16% @ 32.00
      Carpenters - 36% @ 29.00
      Cement Masons - 2% @ 27.10
      Iron Workers & Op.Engrs. - 8% @ 31.50

      Brick Layers - 7% @ 29.60
      Painters - 5% @ 27.80
      Roofers - 5% @ 26.10
      Electricians - 12% @ 33.20
      10 Misc. Trades - 9% @ 28.00

[3] 2 Carpenters @ 29.00 to 1 Laborer @ 22.50 = 26.80

APPENDIX

Construction Industry Statistical Data
2001 (2000 Statistics)
New Construction Put In Place in the United States
(Millions of Dollars)

	PRIVATE[1]		PUBLIC[2]
Residential			
New Housekeeping			
1 Unit	230,000		
2 or More Units	28,500		
Total New Housekeeping		258,500	
Addition & Alterations (Improvements)		63,500	
Total Residential		322,000	
Buildings			
Industrial	30,100		2,000
Commercial, Office	54,700		-
Commercial, Other	59,700		-
Religious	7,900		-
Educational	10,600		27,000
Hospital and Institutional	13,300		5,100
Hotels and Motels	17,100		-
Miscellaneous Buildings	10,100		27,000
Housing and Redevelopment	-		5,300
Total Buildings		203,400	59,070
Heavy Construction			
Public Utilities			
Tele Communications			
Railroads			
Electric Light and Power			
Gas			
Petroleum Pipe Line			
Total Public Utilities			
Sewer			12,300
Water			6,800
Miscellaneous			15,600
Total Heavy Construction			34,700
Highways and Streets			115,200
Miscellaneous			
Farm	2,363		-
Military	-		3,700
Conservation	-		8,000
Total Miscellaneous		2,363	11,700
GRAND TOTAL		574,000	158,000

TOTAL DOLLAR VALUE
TOTAL PRIVATE & PUBLIC CONSTRUCTION 1999   705,000,000,000
TOTAL PRIVATE & PUBLIC CONSTRUCTION 1998   642,000,000,000
1998 INCREASE   63,000,000,000

PHYSICAL VOLUME INCREASE   +7.0%

---

[1] U.S. Department of Commerce
[2] *Not available since Construction Review ceased publication in 1998. 2000 Statistics Estimated.*

APPENDIX

Construction Industry Statistical Data
2001 (2000 Statistics)
Number of Construction Workers by Trade Designation

Workers (Tradesmen)	Number of Employees [1]	Percent of Workers [2]
Basic Trades		
Bricklayers	257,400	5.0 %
Carpenters	1,362,400	26.6
Cement Masons	113,250	2.2
Iron Workers	133,800	2.6
Laborers	1,029,600	20.0
Operating Engineers	164,700	3.2
Truck Drivers	30,900	.6
Specialty Trades		
Carpet and Resilient Floor Layers	164,700	3.2
Elevator Constructors	20,600	.4
Glaziers	25,700	.5
Lathers	10,300	.2
Painters	175,900	3.7
Plasterers	20,600	.4
Roofers and Sheet Metal	179,700	4.3
Terrazzo Workers	10,300	.2
Tile Setters	46,300	.9
Acoustical Tile Setters	36,300	.7
Mechanical and Electrical Trades		
Electricians	677,100	11.4
Plumbers and Pipe Fitters	688,800	11.9
Sheet Metal Workers	100,300	2.0
Total Workers	**5,248,650**	**100.00%**

Architects, Engineers and Contractors		Percent of Total
Architects and Engineers	544,000	38.00
Construction Managers and Estimators	576,000	40.00
Accountants and Clerical	320,000	22.00
Total	**1,440,000**	**100.00%**

All Workers Total	**6,688,650**	

[1] *Construction Review 1998* - U.S. Department of Commerce & AC&E Projections for 2000 Statistics
[2] Percentages - AC&E

Note: Metro Weighted Wage Rate Averages and Modifiers computed from this table of approximate number of active Tradesmen in the industry.

APPENDIX

U.S. Measurements & Metric Equivalents - (For Construction)

Weights & Measures (U.S. System)

Length	1 Mile = 320 Rods = 1,760 Yards = 5,280 Feet
Area	1 Mile = 640 Acres = 102,400 Sq. Rods = 3,097,600 Sq. Yards
Area	1 Acre = 160 Sq. Rods = 4,840 Sq. Yards = 43,560 Sq. Feet
Volume	1 Cu. Yard = 27 Cu. Feet = .211 Cord
Capacity	1 Gallon = 4 Quarts = 8 Pints = .13337 Cu. Feet
Weight & Mass	1 Ton = 2,000 Pounds = 32,000 Ounces

U.S. to Metric - Area, Length, Volume & Mass Conversion Factors

Quantity	From U.S.	To Metric	Multiply By
Length	mile	km	1.609 344
	yard	m	0.914 4
	foot	m	0.304 8
	inch	mm	25.4
Area	square mile	km^2	2.590 00
	acre	m^2	4 046.87
	square yard	m^2	0.836 127
	square foot	m^2	0.092 903
	square inch	mm^2	645.16
Volume	cubic yard	m^3	0.764 555
	cubic foot	m^3	0.028 316 85
Mass-(Weight)	lb	kg	0.453
Mass-(Length)	plf	kg/m	1.488
Mass-(Square)	psf	kg/m^2	4.882
Mass-(Cube)	pcf	kg/m^3	16.018

Length Conversion Tables from Fractions to Decimals to Millimeters

Fraction	Decimal	Millimeters	Fraction	Decimal	Millimeters
1/16"	.0625	1.588	9/16"	.5625	14.288
1/8"	.125	3.175	5/8"	.625	15.875
3/16"	.1875	4.763	11/16"	.6874	17.463
1/4"	.250	6.350	3/4"	.750	19.050
5/16"	.3125	7.938	13/16"	.8125	20.638
3/8"	.375	9.525	7/8"	.875	22.225
7/16"	.437	11.113	15/16"	.9375	23.813
1/2"	.500	12.700	1"	1.000	25.400

Weights - Construction Products

Substance	Wt.Lb. per Cu.Ft.	Substance	Wt.Lb. per Cu.Ft.	Substance	Wt.Lb. per Cu.Ft.
Aluminum	165	Lead	710	Concrete	144
Brass	534	Paper	58	Cement	90
Bronze	509	Glass	156	Clay	100
Copper	556	Granite	165	Sand	100
Iron	450	Limestone	160	Asphalt	81
Steel	490	Brick	140	Wood	40

Computation of Area

Square or Rectangular	=	Length x Width or Height
Triangle	=	Base x ½ Height
Circumference of Circle	=	Diameter x 3.1416
Area of Circle	=	Radius Squared x 3.1416
Cube of a Column or Cylinder	=	Radius Squared x 3.1416 x Depth
Cube of a Solid	=	Length x Width x Depth
Area of Ellipse	=	½ Length x ½ Height x 3.1416

APPENDIX

UNITED STATES INFLATION RATES
July 1, 1970 to June 30, 2002
AC&E PUBLISHING CO. INDEX

Start July 1

YEAR	AVERAGE %	LABOR %	MATERIAL %	CUMULATIVE %
1970-71	8	8	5	0
1971-72	11	15	9	11
1972-73	7	5	10	18
1973-74	13	9	17	31
1974-75	13	10	16	44
1975-76	8	7	15	52
1976-77	10	5	15	62
1977-78	10	5	13	72
1978-79	7	6	8	79
1979-81	10	5	13	89
1980-81	10	11	9	99
1981-82	10	11	10	109
1982-83	5	9	3	114
1983-84	4	5	3	118
1984-85	3	3	3	121
1985-86	3	2	3	124
1986-87	3	3	3	127
1987-88	3	3	4	130
1988-89	4	3	4	134
1989-90	3	3	3	137
1990-91	3	3	4	140
1991-92	4	4	4	144
1992-93	4	3	5	148
1993-94	4	3	5	152
1994-95	4	3	6	156
1995-96	4	3	6	160
1996-97	3	3	3	163
1997-98	4	3	4	167
1998-99	3	3	2	170
1999-2000	3	2	3	173
2000-2001	4	4	4	177
2001-2002	3	4	3	180
2002-2003	4	5	3	184

INDEX

INDEX

A

Abrasive - nosings	302.35
Access doors	1001
Access floors	1305
Accordion doors	1020
Acoustical tile	905
Admixtures - concrete	301.32
- masonry	409.5
Automotive equipment	1115
Air conditioning	1507
Altar	1111
Anchor - bolts	502.1
- slots	302.32
Anechoic chambers	1316.1
Antenna systems	1607.1
Appliances - residential	1106
Ashlar stone	407.2
Asphalt - expansion joint	301.35
- floor tile	908.1
- pavement	206.1
- shingles	707.1
Astronomy observation domes	1315
Athletic equipment - indoor	1107
- outdoor	208.1
Audiometric rooms	1316.2
Auger holes	202.2
Automatic door openers	808.3

B

Backfill	202.3
Backstops - indoor basketball	1107.1
- outdoor basketball	208.2
Bank equipment	1113
Bar joists	504
Base - ceramic tile	903.1
- concrete	301.22
- quarry tile	903.2
- resilient	908.5
- terrazzo	904.1
- wood	603.4
Baseboard heating - electric	1608.2
- hot water	1508.5
Bathroom accessories	1026
Bathtub enclosures	1027
Beams - formwork	302.14
- grade	302.12
- laminated	604
- precast	305.10
- steel	501.1
- wood	604.2
Benches - exterior	208.3
Blackboards	1002
Bleachers - exterior	208.3
- interiors	1107.2
Block - concrete	402
- glass	404
- gypsum	4405
Blocking - carpentry	601.31
Boards - chalk	1002
- directory	1006
- dock	1008
Boilers	1506.4
Booths - parking	1115.3
- telephone	1025
Bowling alleys	1302
Box - mail	1018
Brick - masonry	401
Bridging	601.13
Bucks - doors	601.31
Building paper	601.34
Built-up roofing	705.1
Bulletin boards	1006
Bumpers - dock	1008
Burglar alarm system	1607.3
Bus duct	1601
Bush hammer concrete	301.25

C

Cabinets - hospital	1101
- kitchen	1103
- laboratory	1104.2
- school	1102
- shower	1027
Cable - tray	1601.9
Caissons	203.4
Cants - roof	601.33
Carborundum	301.31
Carpentry - finish	602
- rough	601
Carpet	909
Castings	502.2
Cast iron - manhole covers	205.3
Catch basins	205.3
Caulking	710
Cellular deck - metal	505.3
Cement - colors	301.33
- grout	301.36
- Portland	409.1
Cementitious decks	307
Ceramic tile	903.1
Chain link fence	207.4
Chairs	1205.2
Chalkboard	1002
Chamfer strip	302.33
Chiller - water	1507.3
Chimney - concrete	1303.1
- metal	1303.2
- radial brick	1303.1
Church equipment	1111
Chutes - linen & rubbish	1003
Circuit breakers	1604.5
Clay facing tile	403
Clean rooms	1304
Clear and grub	201.2
Clock & program system	1607.4
Coat & hat racks	1004
Coiling partitions - metal & wood	1020
Columns - forms	302.13
Compaction - soil	202.4
Compactors - waste	1028
Concrete - prestressed	305
Condensing units	1507.2
Conduit - electrical	1601.1
Confessionals	1111.2
Cooling towers	1507.4
Core drilling	201.3
Cork - flooring	908.2
- tackboard	1002
Corner guards - steel	1029
Counter tops	603.2
Courts - tennis	208.1
Cranes	1406
Cribbing	204.6
Culverts	205.2
Curbs - asphalt & concrete	206.2
Curing - concrete	301.34
Curtain - walls	809
- cubicles & tracks	1005

Page 1

INDEX

D

Dampproofing	702
Darkroom - prefab	1004.2
Deck - laminated wood	604.3
- metal	505
Demolition	201
Dental equipment	1104.17
Detention equipment	1114
Dewatering	205.5
Dimmer switch	1601.5
Directory boards	1006
Display & Trophy cases	1007
Disposal excavation	202.5
Dock equipment - loading	1116
Doctors register	15
Domes - skyroof	709.2
Door - blast	804.1
- frames, metal	804.1
- frames, wood	801
- hardware	803
- flexible	804.3
- folding	1020
- glass, all	804.4
- hangers	804.5
- industrial	804.15
- metal	801
- overhead	804.7
- plastic laminate	804.8
- revolving	804.9
- rolling	804.10
- shower	804.11
- sliding	804.12
- traffic	804.14
- vault	804.16
- weather stripping	811
- wood	803
Dovetail slot	302.32
Drywall - gypsum	602.24 and 902
Ducts - electrical	1601.6
Dumbwaiters	1403

E

Ecclesiastical equipment	1111
Elastomeric roofing	706
Elevators	1401
Epoxy flooring	301.25
Escalators	1402
Evaporators	1507.6
Excavation	202.2
Exit lights	1605.1
Expansion joints - concrete	301.35
- covers	1009

F

Fertilizing	209.2
Film viewer, X-ray	1104.10
Finish - hardware	810
Finishing - concrete	301.2
Fire alarm system	1607.2
Fire - brick	401.4
- doors	804.6
- escapes	502.4
- extinguishers	1010
- hydrants	1505.3
Fireplace accessories	412.4
Fixtures - electrical	1605.1
- plumbing	1504.4
Flagpoles	1011
Flagstone	407.3
Flashing	708.3
Flue lining	406.2
Folding doors & partitions	1020
Folding gates	1014
Food service equipment	1108
Fuel tanks	1501.8
Fuel handling equipment	1506.1
Fume hoods	1104.4

G

Generators - emergency	1602.1
Glass and glazing	812
Glazed - block	402.7
- brick	401.2
Glue laminated	604
Grading - site	202.1
Grandstands	1306.1
Granite - building	407.13
Greenhouses	1307
Grilles & louvers	1016
Grounding	1604.2
Grounds	601.32
Guard rails	207.4
Gunite	304.4
Gutters & downspouts	708.1
Gymnastics equipment	1107.7
Gypsum - deck	307.2
- drywall	602.24 and 902
- lath	901.12

H

Hand excavation	202.2
Hardboard	602.13
Hardeners - concrete	301.37
Hardware - drapery	1204
- finish	810
- rough	607
Hatches - floors & roof	1012
Heating	1506
Hollow metal doors	801
Hollow metal toilet partitions	1027
Hospital equipment	1104
Incinerators	1308
Insulated rooms	1309
Insulation - building	703
- pipe & duct	1501.9
Integrated ceiling	1509.4
Intercom systems	1607.5

J

Joints - control & expansion	301.35
Joists - steel	504
- wood	601.12

K

Kitchen - equipment	1108

L

Laboratory equipment	1104.2
Landscaping	209
Lathing	901.1
Laundry chutes	1003
Laundry equipment	1109
Lavatories	1504.4
Lead - shielding	1104.10

INDEX

Letter & signs 1022
Lift slab 304.1
Lifts - garage 1404.2
 - stage 1404.3
 - man 1404.4
Light gage framing 507
Lighting 1605
Lightning protection 1606
Limestone 407.11
Linen chutes 1003
Lintels - steel 502.10
Loading dock equipment 1116
Lockers 1015
Locksets 810
Louvers 1016

M

Mail chutes & boxes 1018
Manholes 205.3
Map rail 1002
Marble 407.12
Marble veneer 906.2
Masonry - restoration 414
Medical equipment 1104
Medicine cabinets 1027
Mercury vapor lights 1605.2
Mesh - partitions 1017
Metal - buildings 1310
 - lath 901.11
 - partitions 1019
Metering - mechanical 1502.6
 - electrical 1604.1
Mirrors 812.12
Miscellaneous metals 502
Monorails 1406
Mortar 409
Movable partitions 1019
Moving buildings 201.1
Moving stairs & walks 1402

N

Nosings - stairs 302.34

O

Oil storage tanks 1501.8
Open web joists 504
Ornamental metals 503

P

Painting - building 912
Pan forms 302.15
 - stairs 502.12
Paneling 502.23
Panels - asbestos cement 809.3
 - concrete 306.3
Parget flooring 910.2
Parking equipment 1115
Partitions - acoustical 1019
 - drywall 902
 - movable 1019
 - toilet 1027
Pavers - floor 906.4
Pedestal floors 1304
Pegboard 602.22
Pews - church 1111.3
Piling - bearing 203
 - sheet 204.1
Pipe & fittings 1501.1
Placing concrete 301.1
Plaques 1022

Plaster 901.2
Plastic - pipe 1501.1
 - skylights 709.1
Plumbing 1502
Pneumatic tube systems 1405
Poles - electrical 1601
Polyethylene film 301.39
Pool equipment 1504.5
Post-tension concrete 304.2
Powered scaffolding 1408
Precast - concrete 306
 - panels 408
Pre-engineered buildings 1310
Prestressed concrete 305
Prison equipment 1114
Projection screens 1203
Protected metal 506
Public address & paging system 1607.6
Pumps - condensate 1501.5
 - heat 1501.5
 - water 1504
 - well 1502
 - wellpoint 205.2

Q

Quarry tile 903.2

R

Radiation protection 1311
Rafters - wood 601.14
Railroad work 212
Ranges - shooting 1107.8
Receptors - shower 1021
Refrigeration 1507
Reinforcing - concrete 303.1
Residential - appliances 1106
Resilient flooring 908
Restoration - masonry 4
Rip-rap 204.5
Roads & walks 206
Roof decks - insulation 704
 - metal 505
Roofing - built up 705.1
 - skylights 709.1
 - traffic 706
 - ventilators 709
Rubber - flooring 908.4
 - roofing 706
Rubbish chutes 1003
Running tracks 208.1

S

Sandblasting - walls 414.2
Saunas 1312
Scaffolding 416
Scales 1313
School equipment 1105
Scoreboards 1107.3
Screeds 301.2
Screens - projection 1203
 - urinal 1027
 - window 808.1
Sealants 710
Seamless flooring 907.1
Seating 1207
Seeding & sodding 209.1
Septic tanks 1503.4
Service station equipment 1117
Sewage treatment 1503.4
Shades 1202
Sheathing 601.16

INDEX

Sheetrock 902
Shelters - loading dock 1116.4
Shelving - library 1110
    - storage 1023
Shingles - asphalt 707.1
    - asbestos 707.2
    - wood 701.3
    - slate 701.5
    - tile 701.6
Shoring 302.2
Siding - metal 602
    - wood 602.1
Signs 302.2
Site drainage 20.5
Skydomes 1315
Slate - chalkboard 1002
    - panels 906.3
    - roofing 707.6
Slides - playground 208.2
Snow melting equipment 1608.1
Soffits 602.17
Soil - compaction 202.4
    - poisoning 210.1
    - testing 202.6
Soundproof enclosures 1025
Sound insulated room 1316
Sound & vibration control 1317
Spiral stairs 502.13
Sprinklers - building 1505.1
Stage & theatrical equipment 1112
Steel - buildings 1310
    - studs 902.4
Sterilizers 1104.9
Stone 407
Stone - flagging & paving 906.12
Stone veneer 906
Structural steel 501
Stucco 901.24
Studs - metal wall 902.4
    - wood 601.11
Substations 1603.1
Sun control devices 1024
Swimming pools 1318
Swing - playground 208.12
Switchgear 1601.4
Switchboards 1604.4

T

Tackboards 1002
Tanks - oil, gasoline & water 1501.8
Telephone booths 1025
Telephone equipment 1607.1
Television antennae 1607.7
Temperature control equipment 1609.3
Tennis courts 208.1
Termite - soil treatment 210.1
Terra cotta 408.2
Terrazzo 904
Ties - form 302.31
Tile - acoustical 905
    - ceramic 90
    - plastic 913.2
    - quarry 903.2
    - resilient 908
    - roofing 707.6
Tilt up construction 304.3
Timber - framing 601.2
Toilet - accessories 1026
    - partitions 1027

Tops & bowls 1104.3
Tracks - running 208.1
Transformers 1604.6
Trash chutes 1003
Trees & shrubs 209.5
Trench covers 502
Truck - docks 1404.1
    - scales 1313
Trusses - wood 605.1

U

Underpinning 204.4
Utilities - sitework 205

V

Valves 1501.4
Venetian blinds 1202
Vibroflotation 210.3
Vinyl asbestos flooring 908.6
Vinyl flooring 908.5

W

Walks 206.3
Wall coverings 913
Wall urn 1030
Waterproofing 701
Water stops 302.35
Water treatment 1502.4
Windows 805
Wire - electric 1601.3
    - mesh 303.2
Wood - doors 803
    - flooring 910
    - framing 601.1
    - piling 204.2
    - treatment 606
    - trusses 605.1

X

X-ray equipment 1104.1